Space-time for Absolute Beginners

Tony Goldsmith

First published by Tony Goldsmith in the UK

Copyright © Tony Goldsmith 2018
Volume 1 of Absolute Beginners Series

The right of Tony Goldsmith to be identified as the author of this work has been asserted by him in accordance with the Copyright, Designs and Patents Act 1988.

All rights reserved. This book is sold subject to the condition that it shall not, by way of trade or otherwise, be lent, re-sold, hired or otherwise circulated in any form of binding or cover.

This book has been revised (most recently on 16/7/18) and renamed to complement my second book which is Volume 2 of Absolute Beginners Series called 'Quantum Physics for Absolute Beginners'. The original paperback called 'How to Win a Nobel Prize' has been discontinued.

The Kindle version of 'How to Win a Noble Prize' currently retains all of the original narrative.

Contents

    Preface
1. What is Real and what is Perceived?
2. Background Information
3. What do we know about Space?
4. What do we know about Time?
5. What is the effect of Gravity?
6. How did our Universe Begin?
7. How did we get Here from There?
8. What was the Space-time Catalyst?
9. What is Space-time?
10. How does Space-time work?
11. How to win a Nobel Prize?
    Conclusion
    Notes
    Holes in the Truth
    Further Reading
    Index
    About the Author

Preface

In the Twentieth Century Albert Einstein and others made some astonishing discoveries. They showed us how our universe could start with a Big Bang which seems to allow something to appear from nothing, that Space can be curved and why there was a time when there was no Time.

If you have always been absolutely baffled by anything related to Space-time then this book is for you. My writing mission is to engage with a new audience that has no previous knowledge of science, and I go to great lengths to make sure that everybody can fully understand my explanations.

Unfortunately, Space-time has a reputation for being difficult. The first thing I will do is show that this claim is false. In this book nothing is too hard to understand and I will also encourage you to participate in the scientific process.

An important aspect of this book is the ease with which the reader can follow my scientific explanations. I will make use of bullets when a topic is complex or has a logical sequence. For the most direct read you can ignore all brackets. Anything in brackets is only there if I think the text will benefit from some clarification (or to give you the chance to google an **emboldened** word before you carry on). If you ignore the brackets then you will not encounter any jargon within the text.

I will ignore any detail which (I believe) does not assist with your initial understanding of this subject. Any obvious omissions are listed in 'Holes in the Truth' at the back of the book and anything that requires mathematics or scientific jargon is relegated to the 'Notes' section.

In the last 30 years there have been no big breakthroughs in our knowledge of physics and the world that we live in. Despite the media attention surrounding the detection of Gravitational Waves, the exposure of the Higgs Boson at CERN and the Hubble space telescope evidence for the expansion of the universe twenty years earlier, all of these discoveries were merely confirmations of predictions that had been mainstream scientific theories for over 50 years.

Einstein was not originally a career scientist; he worked in the Swiss Patents Office. His thought experiments were based upon posing questions that no scientist had previously asked. What is needed right now is a good question to enable us to make a breakthrough in the Twenty First Century.

Scientists are struggling to achieve this so it is my contention that the time has come for people not constrained by previous thinking to make their contribution and I will also encourage your participation in the scientific process. Don't worry, you are not going to need your own laboratory; everything can be done in the comfort of your own home.

1   What is Real and what is Perceived?

In this book we are going to take a journey through the development of macro physics knowledge. We will start in this chapter by ironing out some definitions.

You will already have your own interpretation about the definitions of Real and Perceived. However, in this book I am going to use these words as the foundation for some new ideas which I will develop throughout the book.

1.   What do I mean by Real and Perceived?

You will be familiar with sunrise and sunset (and hopefully you will have seen some nice ones). However, these are not scientific terms. They are events that 'emerge' due to the rotation of our Earth. I cannot convince you that they do not exist but they will not assist in a quest to find the essence of reality. They are not fundamental.

In 'Quantum Physics for Absolute Beginners' I analysed the validity of using Matter and Light as useful concepts within reality. Clearly, I cannot convince you that Matter and Light do not exist but, to understand what is fundamental in nature we have to dig deeper.

It is Quantum Physics that is the concept which is Real with Matter and Light relegated to items which are Perceived.

Throughout this book I will insert a diagram whenever it is likely to be useful. My view is that one diagram is always worth at least a thousand words.

MATTER   LIGHT

PERCEIVED

↑   ↑   ↑

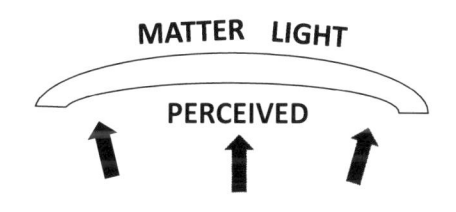

QUANTUM PHYSICS

REAL

You may think that you know all about Matter and Light but if you do not know what Quantum Physics is then I am afraid that you are going to have to relearn everything that you thought you knew. Matter and Light do not exist separately. They are partners in a famous double act:

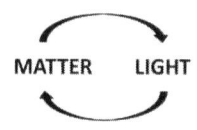

MATTER   LIGHT

There are different sorts of double acts. In your life you will have encountered many sorts of double acts. There are superheroes like Batman and Robin, dancing duos like Fred and Ginger and comedy couples like Laurel and Hardy. Clearly, these are all different.

In this book I will stop every time I find a new double act. Matter and Light belong to the group that cannot live apart.

We are now ready to ask some new questions:

2. **What happens if Space and Time are not Real but only Perceived:**

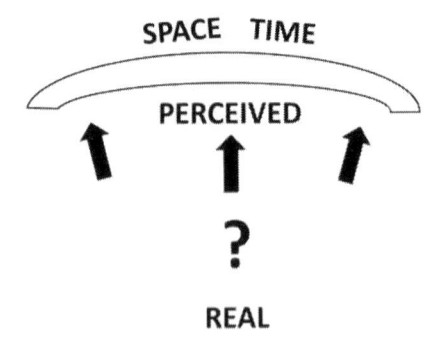

```
        SPACE   TIME
      ⌒⎯⎯⎯⎯⎯⎯⎯⌒
          PERCEIVED
        ↑    ↑    ↑

             ?

           REAL
```

This time the 'Real -> Perceived' diagram is not completed. I have some answers ready to discuss with you and we will start to take a look at these soon. However, I don't know all of the answers. To appreciate a full understanding of Space and Time we need to find answers to the following three related questions:

1.	If Space and Time are Perceived then what is Real?
2.	What changed to make it Real?
3.	When did this happen?

This book will provide you with some of the most important ideas about the potential answers to these three questions. In the final chapter you will get the opportunity to add any new ideas that you might have (by then).

As you will soon see, the hero of our story is Albert Einstein. He very rarely entered a science laboratory and he was not brilliant at mathematics. He was very good, however at asking the right questions and formulating Thought Experiments which gave new insight into the true nature of reality.

Science is all about asking new Questions.

2 Background Information

The Space-time story will take a few twists and turns before we reach the end of our journey. This chapter covers some basic concepts which we need to clarify before we get into the nitty-gritty.

1. What is Matter?

This question is covered in depth in my book 'Quantum Physics for Absolute Beginners'. For the purposes of this book you just need to know:

An atom is the smallest chunk of matter that can exist without losing its identity. It has essential building blocks as shown below:

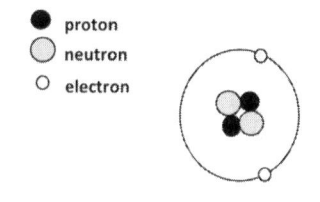

Helium Atom

The figure above is the High School model of a Helium atom. It has a nucleus of 2 protons and 2 neutrons and it has 2 electrons in its inner energy level. As you can immediately see it is very symmetrical.

2. What is Light?

Light is the entity that allows us to see things. It travels at approximately 300,000 kilometres per second (km/sec) in air. That is about ¾ of the distance to the moon in one second. We are talking faster than anything you can imagine!

The speed of light has its own abbreviation which is **'c'**. Light consists of photons which are depicted in diagrams as wavy lines.

Photon

3. What is a Photon?

A wave of light is actually a convoy of photons. Our universe is awash with photons which are either travelling somewhere or colliding with an atom. For a full analysis of photon interactions with atoms see Volume 2 of this Series.

For the purposes of this book you just need to know that photons spend their whole lives travelling between atoms.

In a vacuum photons are rushing in every direction during every moment without colliding with each other or losing any information. This phenomenon is very different to anything else we know about.

4. What is Electromagnetic Radiation

There are seven categories of photons which, when combined, make up the Electromagnetic Radiation Spectrum.

The main thing that makes photons different is their energies.

I have grouped all Electromagnetic Radiation into categories as shown in the figure below:

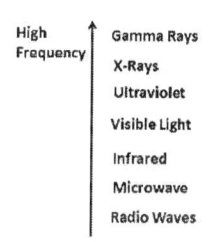

```
High  ↑   Gamma Rays
Frequency X-Rays
          Ultraviolet
          Visible Light
          Infrared
          Microwave
          Radio Waves
```

Electromagnetic Spectrum

a. The photons that we see with our eyes are in the Visible Light frequency section – this is tiny compared to the whole spectrum
b. Infrared radiation is the heat radiation that comes from a hot source and is clearly illustrated by an electric bar fire
c. Microwave radiation is used for satellite dishes and broadband routers (as well as microwave ovens)
d. Radio waves are used for transistor radios and terrestrial television aerials
e. All of the categories mentioned so far are everywhere all of the time; wherever you take your transistor radio or your laptop computer you expect to get a signal
f. It is just as well that the electromagnetic radiation above the Visible Light section is not everywhere all of the time. If it was then we would all be dead because this radiation has lots more energy
g. X-Rays and Gamma Rays are high energy (and therefore high frequency) radiation which we need to avoid as much as possible.

5. What is a Field?

The word 'field' was used for the first time in 1845 by Michael Faraday. Scientists needed something to explain mysterious action at a distance.

The most obvious field can be displayed by scattering iron filings around a bar magnet. When you do this you see the picture below on the left:

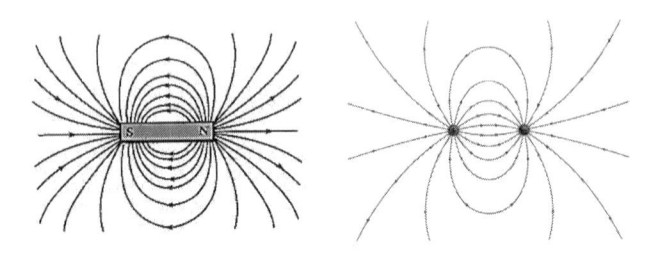

Bar Magnet Magnetic Field and Electric Field

It is a picture of the magnetic field. There is a similar field around electric charges (shown on the right) but this is harder to demonstrate in the lab.

For all fields, where the lines are closest (i.e. near the 'source') the field is strongest. Further away the field gets weaker and eventually 'sinks'. All fields present themselves in this way.

In the twenty-first century we are now aware of many fields. You will have known about Gravitational Fields and you may have heard about Gravitational Waves moving through it.

6. How do Photons move?

Michael Faraday also showed that a moving electric field would induce a magnetic field and a moving magnetic field would induce an electric field.

In 1865 James Clerk Maxwell wondered what would happen if a magnetic field induced an electric field and it then induced another magnetic field and so on. If you could do this you would have found a perpetual motion machine.

Maxwell worked out how fast this would have to happen and it was near enough 300,000 km/sec. Not only had Maxwell calculated the most accurate value for the speed of light but he had also shown that light (aka **electromagnetic radiation**) comprised of two fluctuating fields.

We now know that these fluctuating fields overlap each other so that the fields never cease to exist. This can be done with a spiralling motion as shown below:

Electromagnetic Propagation

Faraday had done all of the experimental 'spade' work but it was Maxwell's brilliant insight which discovered the incredible fact that it is two fields overlapping that provides us with light and all the amazing views we have of our world and our universe.

In science we often stumble across double acts. When this happens I will comment and highlight it with a diagram. If you think that I miss any then please write them down and retrieve them when we get to the final chapters.

The next double act is an Electric Field and a Magnetic Field (aka **equivalence)**:

ELECTRIC   MAGNETIC
  FIELD   FIELD

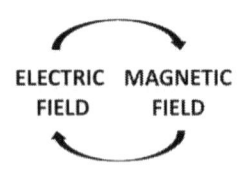

7. Do we need Graphs?

There are different sorts of graph. You may remember those you drew in school. These were probably distance v time graphs. I have shown what these look like below left:

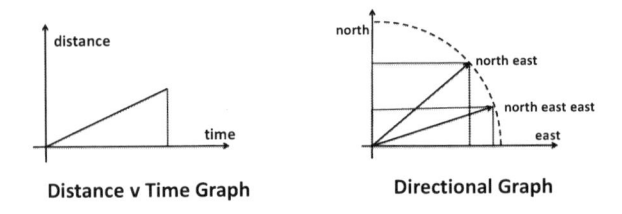

Distance v Time Graph   **Directional Graph**

One axis deals with distance and the other with time. Using these parameters we can calculate things like speed. We are going to take a close look at speed but, don't worry, we have none of these types of graphs in this book.

Instead, we have a few directional graphs. An example is shown top right where we consider movement north and east.

The direction indicator is the same length for each direction. Most importantly, each direction can be split into some northerly direction and some easterly direction (aka **vectors**). Most of my graphs are like this.

As well as a few graphs I will utilise a lot of diagrams. Be careful and do not turn them into graphs in your head.

8. What is Anti-Matter?

In an atom, protons have positive charge and electrons have negative charge. When they live together inside an atom (with the same numbers) the atom is charge neutral. For a full understanding of matter you also need to understand anti-matter.

The best example we have of anti-matter is a positron which has the same mass as an electron but its charge is the opposite. Positrons even occur naturally (as part of **radio-active radiation**) but they don't last long. Other fundamental particles, such as protons, have their anti-matter counterparts too.

All forms of anti-matter disappear (**annihilate**) when they come into contact with ordinary matter (i.e. the stuff all around us). Their combined masses dissolve into energy which is released as electromagnetic radiation. All other features cancel each other out (i.e. **charge**).

Anti-matter can be created in labs and it can be brought together to form anti-atoms. It has been 'manufactured' at the home of the European Organisation for Nuclear research (based in **CERN**) but it can only be stored in expensive force fields stopping it from touching the side of a container.

9. Where does Anti-Matter come from?

There is a proverb that states 'Space abhors a vacuum'. It used to mean that it is very hard (virtually impossible) to construct a vacuum on earth and, if you do, it is just as hard to preserve it. This is true on a universal scale because we now know that a cubic metre of 'vacuum space' (far away from any planetary atmosphere) is not empty at all.

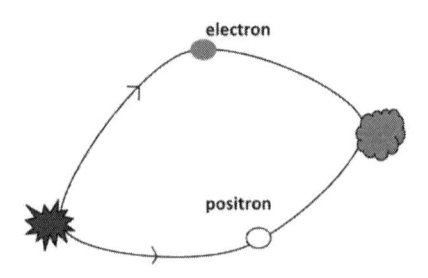

Vacuum Virtual Particles

As shown in the figure above Vacuum Space allows something (**virtual particles)** to constantly appear and disappear. These particles will normally have the same mass but opposite charge (such as an electron and - its anti-matter twin - a positron) and will only exist until they recombine and annihilate each other. This happens too fast for us to view it with our measuring instruments but we can detect the effect from other experiments (see **Casimir Effect**).

Amazingly, this constant dance of matter and anti-matter is happening everywhere all of the time. This is another double act where both entities are similar:

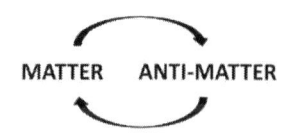

MATTER ⇄ ANTI-MATTER

In Chapter 2 we have summarised the microscopic view of science. You may have already known something about atoms and, if so, take that as your revision for the day. At this point it is just important to know that matter is made of atoms and light is made of photons. If you did not know anything about atoms and light you do now and, most importantly, we are all now on the same page.

Now, let's go large.

3 What do we know about Space?

From your perspective space and time just carry on, day in and day out – this is what our limited senses tell us. You are in good company because Isaac Newton and most other scientists up until 1905 would agree with you. Newton's Laws explicitly assume that space and time are continuous and the same everywhere; thereafter Newton ignores space and time and concentrates on forces, mass and acceleration etc. to give us his famous Laws (see Notes 1).

The implication is that space and time are the same for everybody. This suggested that they are experienced the same by everybody and are just something sitting in the background whilst mainstream physics with forces and fields take centre stage in the foreground.

In 1905 Einstein changed all of that. He published a paper that became known as the 'Theory of Special Relativity' and the scientific world would never be the same again. Whatever we thought we knew about space and time, it was now time to think again and this is what you are going to have to do in this chapter.

Most of Einstein's peers would not accept these new radical claims and he struggled for decades to achieve scientific acceptance (it was a Big Change). Arguably there have been four major Eureka discoveries in Physics and this was one of them. (The others were **Newton's Laws**, **Maxwell's** equations for **Electromagnetic Radiation** and **Quantum Mechanics** which was effectively revealed by a 'committee' of Scientists.)

1. What is it like moving at constant speed in space?

In our world we are all familiar with cars and airplanes and we think we understand what speed is. However, that force that pushes you into your seat is not speed. It is acceleration and we will look at this in Chapter 5. Many things move at constant speed and we are all doing it right now.

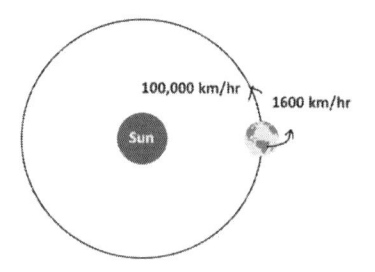

Constant Speed

a. The Earth is turning round on its axis at about 1600 kilometres per hour (at the equator) and it orbits the Sun at about 100,000 km/hr as shown above.
b. If we could feel this movement we would be forever falling over whilst walking down the road.
c. If you were in a rocket and travelling at constant speed, far away from any planet with gravity, then you would not be able to know if you were stationary (i.e. standing still on Earth) or moving at constant speed.

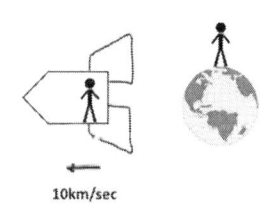

Constant Speed v Standing Still

d. As shown in the diagram above, according to the Laws of Physics, from your viewpoint moving at constant speed and standing still are the same (i.e. **equivalent**).

e. If this was not the case and standing still was somehow special then we would need a reference point to know we were stationary.
f. What could that reference point be? The Sun or the centre of our Galaxy? All of these things are moving too!
g. We now understand that there is no reference point - standing still and moving at constant speed are equivalent.
h. This is a motion double act:

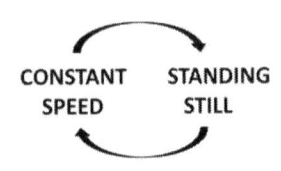

CONSTANT SPEED ⇄ STANDING STILL

This is one of those laws learnt through experiment and logic.

2. What is Classical Relativity?

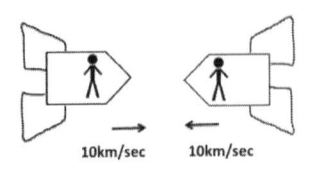

10km/sec   10km/sec

Rockets Approaching

a. In the figure above we see two rockets approaching each other. Let's suppose that each rocket is travelling at a constant speed of 10 km/sec. This is a Thought Experiment so we can set the speed to anything we want. If they collide the impact will be the equivalent of a 20 km/sec crash. Within our everyday world you are well aware that the size of a crash on

the road is dependent upon not only your speed but also the approaching car (whose fault it was!).
b. This is known as 'Classical Relativity' which was first documented by Galileo. He did not have a fast car so he had to work this out through a Thought Experiment and logic.
c. Classical Relativity states that the speed of each rocket relative to the other is 20 km/sec. We can imagine that one rocket man is standing still and that the other rocket is moving at 20 km / sec relative to this position. Alternatively, the second rocket could be standing still and the other would be moving at 20 Km /sec.
d. Each is relative to the other because there is no absolute reference point (for standing still).

3. What is the speed of a bullet fired from a rocket?

The diagram below shows a rocket approaching Earth.

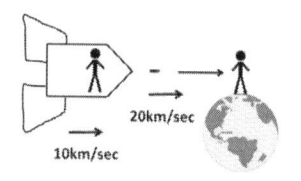

Rocket shooting at Earth

a. The rocket is approaching at 10 km/sec and we are observing this on Earth. According to Classical Relativity both the rocket and the Earth are approaching each other at 10 km/sec. All movement is relative to the position you take.
b. Let's make it simpler and declare that the Earth is stationary so it is just the rocket moving at constant speed and we will firstly look at something with which you are familiar – a bullet.

c. The rocket fires a bullet at the man on the Earth. He is able to measure its speed as it whizzes by and you will expect its speed to be 30 km/sec (i.e. the speed of the rocket plus the speed of the bullet) and it is.
d. For the record the bullet thinks the Earth is whizzing by at 30 km/sec (but that's not important here). This is what Relativity means though.

4. What is the speed of light approaching from a rocket?

This simple question is probably the best one in this book. If you do not already know the answer, prepare to be amazed:

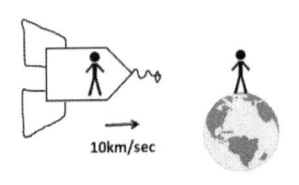

Rocket communicating with Earth

a. Let's have the rocket aim a beam of light (i.e. photons) at the man on Earth
b. As far as the man in the rocket is concerned the light is emitted at the normal Speed of Light which is 300,000 km/s
c. What does the man on Earth measure its speed to be?
d. We would expect it to be the speed of light in air (300,000 km/sec) plus the speed of the rocket emitting the light (10 km/sec) making it 300,010 km/sec (just like the bullet).
e. **But it isn't.** The speed of light remains the same no matter what speed it starts at. It is always 300,000 km/sec no matter what.

This is very different to how we normally perceive our world (i.e. with bullets etc.). It would be the same if the rocket was flying away from the Earth. This is the key fact in Einstein's Theory of Special Relativity which shows that the Speed of Light is not relative at all. Whatever the circumstances, the Speed of Light is Constant!

Einstein recognised this irony and preferred to call it his Theory of Invariance. However, everybody else still calls it the 'Theory of Special Relativity'.

5. So what is Relative about the Theory of Special Relativity?

The Theory of 'Special' Relativity is so-called because it just applies to objects which are at a constant speed. Anything moving at constant speed (like being on a train) feels the same as if we were standing still. This is why, with our eyes shut, we never notice the earth, or the train moving.

So if the speed of light doesn't change what does change? Well speed is governed by the equation - speed equals distance divided by time. Putting that into a 'light' context we have:

Speed of Light = <u>distance travelled by a photon through SPACE</u>
TIME taken

If the Speed of Light stays the same then something strange must be happening to each person's experience of distance (through SPACE) and TIME.

Both the men (one in the rocket and one on Earth) observe the same event but each sees the photon travel a slightly different distance over different time durations. This is because they are in different places moving at different speeds (i.e. different **reference frames**). They both observe and measure distance (through SPACE) and TIME differently.

This section is really important and central to the book. If you are still having problems accepting this then please go to Notes 2 where I show how this works using some examples.

> **The Speed of Light is always the same so this means that SPACE and TIME must be changing**

Space and Time is another double act where the entities are inseparable (like Matter and Light):

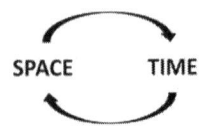

SPACE ⇄ TIME

Einstein showed in 1905 that Space and Time are experienced differently by people in different circumstances. Einstein's peers refused to accept this and consequently he never received the Nobel Prize for this work, arguably the biggest 'blunder' in Nobel Prize history.

6. What happens to Space and Time at the Speed of Light?

One of Einstein's early Thought Experiments was to imagine what it would be like to ride on the back of a photon. He did this whilst working in the Swiss Patent Office. Einstein looked out his window and imagined that his photon was travelling away from the clock tower. As he looked back he would not see the light from the clock hands because it could not catch up with him (riding the photon at the Speed of Light). Consequently, he would not see the clock hands move round and it would appear that time had stopped.

Whereas this started out as the obvious outcome of his Thought Experiment, Einstein later realised that this is what really happens and he formulated the Theory of Special Relativity to prove it.

In fact we now know that, for a photon, time (as we know it) does effectively stop. In his new Theory Einstein explained that, as something travels closer to the speed of light its length shortens (**length contraction**) and time slows down (**time dilation**) until it stops. This is relative to the stationary position that we hold.

A common undergraduate exam question used to ask the following:

If a pole vaulter with a 5 metre pole runs towards a shed which is 2.5 metres wide, relative to an interested bystander, how fast must a pole vaulter run to fit himself and his 5 metre pole momentarily in the 2.5 metre open-ended shed (as shown in the figures below):

Approaching the shed   **Moving at 86% c**

The answer is about 86% of the speed of light so you are not likely to witness this on earth (see Notes 3 if you want to check the maths). It is the foundation of our universe, though and if we ignored these effects we would not have made the technological advances that we see around us (see Chapter 10).

7. What about Mass?

As well as showing that length (i.e. Space) and Time change (for observers in different locations) the Einstein equations also showed that there was a close relationship between Energy and Mass. We get the most famous equation:

$$E = mc^2$$

This buy product from the Theory of Special Relativity is the most famous corollary in the history of science and it means that we have another double act:

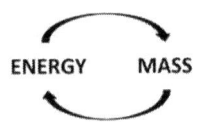

ENERGY   MASS

This relationship is analysed further in Volume 3 of this Series; 'Physics Mysteries for Absolute Beginners'.

As something moves faster and faster (close to the Speed of Light) mass will increase too. All three metrics (length, time and mass) change at the same rate.

8. Is there a Universal Speed Limit?

Utilising his new Theory Einstein showed us that moving very fast (close to the speed of light) makes length contract and slows time down.

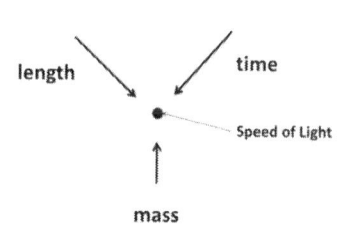

Space Time and Mass near Speed of Light

In the diagram above the Speed of Light is represented by a dot in the centre. As they get closer to the Speed of Light length (Space) and Time will decrease. These are not illusions (i.e. time appearing to go slower); these are physical effects that we can measure (you will really age more slowly on a rocket trip compared to somebody that waits for you on Earth).

We can also calculate the increase in Mass as an object gets closer to the Speed of Light (see Notes 2 for the calculation). As you can see, as these three entities get closer to the Speed of Light something has to give. There is a limit to how small and slow things can get whilst, at the same time, becoming massive. The thing that prevents a physical catastrophe from happening is the Speed Limit imposed upon Light (and its electromagnetic radiation friends). Einstein called this the Universal Speed Limit and, since then we have never seen anything travel quicker than the Speed of Light.

9. **How does a Photon manage to move at the Speed of Light?**

   a. A photon gets over this problem by having zero mass.
   b. It also experiences zero time.
   c. When a photon from a distant galaxy (maybe 10 billion light years away) hits our eye we assume that it has had a long journey. However, as far as the photon is concerned no time has passed at all.
   d. Time is relative which means that different entities experience it in different ways. From the photon's perspective everything is immediate.

10. **Why are we so sure that there is a Universal Speed Limit?**

I have devised my own Thought Experiment:

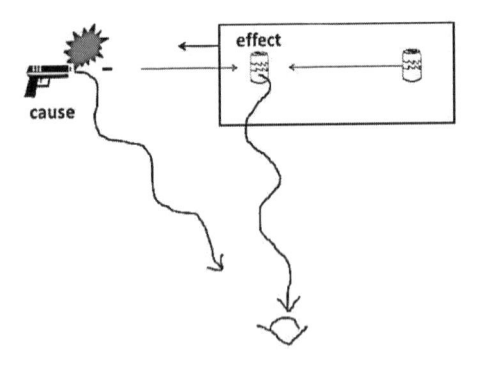

Cause and Effect

   a. A bullet is fired at a can which starts far right and a person views this event. There is a flash of light and the bullet travels from the gun towards the can (in that order) with the person in the middle.
   b. Clearly, we expect to see the flash of the gun before the bullet hits the can. This is the process with which you will be familiar (aka **cause and effect**).

c. Now imagine that we can speed the bullet up to move faster than the Universal Speed Limit and we also move the can towards the gun (on some kind of vehicle) again faster than the Universal Speed Limit so that the bullet and can arrive together almost instantaneously.
d. If these items move fast enough it is possible to see that light from the crumpled can could reach your eye before we see the light from the flash. As you can see above, it now has a shorter distance to travel.
e. Once you have convinced yourself that this is possible (even if the speeds involved are enormous) you must now accept that, within this Thought Experiment, it is possible for effect to precede cause (according to your senses).
f. If you want to insist that effect cannot precede cause then you will have to accept that time is moving backwards in the thought experiment – you will not like that idea any better than the idea (e) above.
g. Because we do not see this in our universe and it is very dangerous to question the validity of cause and effect (i.e. what will you replace it with) we presume that the Universal Speed Limit cannot be broken.

We do not have a definitive Law of Physics (yet) that prohibits movement beyond the Universal Speed Limit and some scientists are still trying to disprove the Universal Speed Limit conjecture (see **Tachyons**). However, clearly the above thought experiments are going to make this very difficult for them.

We now know that Space and Time are experienced differently by people in different circumstances. (Scientists say that Space and Time are not **absolute**).

You are no longer a science fresher and you already know more than many scientists did a hundred years ago. You might even believe it too, which many of them did not.

4   What do we know about Time?

Like space, the early scientists (i.e. those before 1905) assumed that time just carried on regardless. They had no reason to think otherwise. However, we now know that it is reasonable to ask questions like 'What is the direction of time?' In fact, many of those early scientists had the answer to this question but they couldn't put two and two together.

The first issue is that we only have one word for time, but we really need more. The fact that I can use the sentence 'there was a time when there was no Time' (in the Preface) suggests that we have a problem, so let's start there.

1. What do we mean by Time?

Try to answer the question – 'What is Time?' before you continue:

   i. If we state 'Last Saturday at midday' – we have defined a Location in time
  ii. If we state that 'a television programme took 2 hours' – we have defined a Duration of time
 iii. If you say that 'I need to fix the roof ready for the winter' – you have recognised the Flow of time.

In the same way that Eskimos supposedly have 50 words for snow it may well be the case that we need as many to cover all aspects of Time, but in this book we will focus upon these three.

So what is your answer? Most non-scientists, when first asked this question will go for time flow.

All observations in this book so far have referred to time duration but, beware, this is going to change and the other two will take centre stage.

If you have problems understanding new concepts introduced in this book it is probably because you have not come to terms with the fact that Time means different things in different circumstances. The three definitions of Time above are very different. Throughout this book you will find illustrated reminders of the Time definition under discussion.

Time ?

2. Where does Time come from?

1. Good question?
2. What do you mean by time?
3. Is it i, ii or iii above?
4. Are we talking about your time or my time (they are also different)?

3. What can eggs teach us?

You have all broken an egg and, if you haven't, it's high time that you did. The shell smashes into tiny pieces and it allows the white and yoke to escape (into a dish if you are a good cook). What about if you changed your mind about having scrambled eggs; have you ever wondered how hard it would be to put the egg back together again?

Un-break an Egg?

Well, you have. For a long time you have known that this is just not possible. But do you know what is stopping you? Scientists have also known for some time that you cannot 'un-break' an egg. The reason they give is that the world obeys a Law of Disorder (aka the **Second Law of Thermodynamics**) where everything gets more disorderly.

Every day on planet Earth things get more disorderly. If you could 'unbreak' an egg then you would effectively break the Law of Disorder and this is not what we see around us every day.

4. **How do we know we are going forwards in Time?**

Nearly all the Laws of Physics are time independent (i.e. **time reversal** is possible). What do we mean by time independent?

a. The figure below shows a typical game of pool situation just after the cue is taken away. The white hits the black and the black bounces off the cushion.

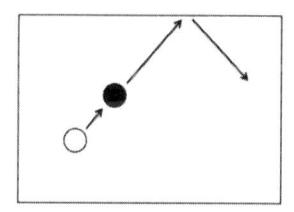

b. If you took a video of the pool balls colliding and then played it backwards (bottom figure) you would not be able to tell, using the Laws of Physics, which was the way forwards (i.e. the direction of time when you made the video).

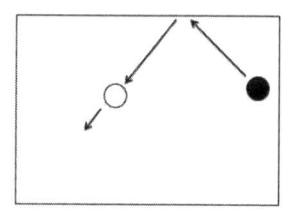

Pool balls colliding

c. In fact, all of the Laws of Physics are time independent (like colliding pool balls) except one.
d. The one law which only works one way in time is the Law of Disorder.
e. The Law of Disorder states that disorder in an <u>Isolated</u> system will always increase or remain static; it will never decrease.
f. We can clearly see this at work in our universe. There is just about nothing you can do in the kitchen which can be undone; try taking the milk out of your coffee or the eggs out of your Victoria sponge.

The Arrow of Time is defined as that dimension moving in the same direction as 'Disorder' which we know goes from past to future.

Thankfully, nobody has <u>seriously</u> suggested that there could be more than one time dimension – that could get really messy!

If you try to argue with any of this then you have to be ready to unscramble eggs!

5. How do we measure Time?

Galileo was the first to see that a pendulum always moves from side to side during the same period of time, irrespective of how long the swing is. This principle was latterly used as the foundation for all clocks up until the mid-twentieth century.

Thereafter, clock makers found that they could use atomic oscillations (using quartz) to provide regular timing events.

You might argue that these machines are measuring time, but they are not. They are simply comparing their measurement of a time period with a different method. You need to come to terms with the fact that Time (itself) cannot be measured. However, this is not the end of this discussion and we will approach it from a different perspective in the next Chapter.

We now need to bring Gravity into the puzzle and see where that takes us.

5 What is the effect of Gravity?

In the Seventeenth Century Isaac Newton showed us why we don't all fall off the planet. His equation for Gravitational Attraction was a brilliant discovery and explained how we could all live on a spherical planet and it revealed how most of the planets move within our Solar System (but not everything).

However, Gravity is by far the weakest force in nature; it cannot even pull a fridge magnet down to Earth. Consequently, when dealing with Gravity we often have to revert to using metaphors.

1. **What happens when we drop a bowling ball onto a trampoline?**

Compared to the trampoline canvass the bowling ball is very heavy and it will stretch the canvass and deform it. The bowling ball will then be propelled upwards as the canvass releases energy and straightens out.

Depending on the forces involved this could continue for some time but eventually the canvass and ball will settle down. At that point when we look at the stationary bowling ball, the trampoline canvass is now permanently curved in a gradual contour.

This is what space really looks like around our Earth except that space has influence in three dimensions and not just the two dimensions of the trampoline sheet shown.

Note that it took a little time for the canvass and ball to settle down into a steady state (i.e. **state of equilibrium**). How long would that take in our Solar System? We are going to look at this question below.

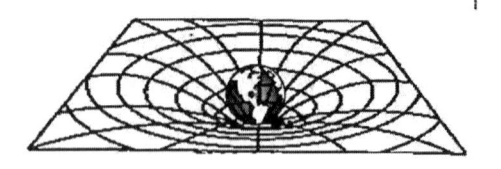

Space around Earth

The diagram above is a 3D slice of space around Earth. Space is curved in every direction and not just the plane shown.

2. What is the effect of Gravity on Space and Time?

Einstein followed his ground-breaking 'Theory of Special Relativity' 10 years later in 1915 with the 'Theory of General Relativity' and this included the effects of Gravity. As always, Einstein started with a Thought Experiment. This one was 'What is the difference between a person standing on the Earth and a person standing in a rocket (in outer space) which is accelerating under the influence of the same force as Earth's gravity?' (See the figure below):

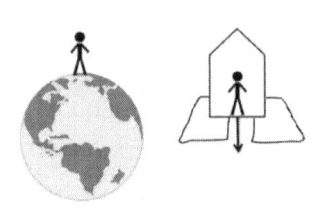

Gravity and Acceleration Equivalence

Firstly, Einstein convinced himself that 'Gravity' and 'Acceleration' are another motion based double act (like Constant Speed and Standing Still).

GRAVITY   ACCELERATION

A person is pushed into the Earth in the same way that a person is pushed into the accelerating rocket.

Einstein then went on to develop the Theory of General Relativity. This tells us that space is curved around mass and that very large mass, like our Sun, makes this visible.

Whilst we might point our moving rocket at a planet (see figure below) the curved space around it will automatically take us on a diversion. We will no longer travel in a straight line.

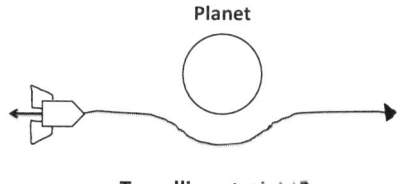

Travelling straight?

Einstein said that these effects occur within a new entity which was called Space-time. The word Space-time was first introduced by Hermann Minkowski in 1909 to explain Special Relativity but, arguably, Einstein was the first to explain it as a new concept.

Large concentrations of Mass curve Space-time and curved Space-time tells Mass how to move.

The amazing implication here is that Gravity is not a separate force but the result of space curvature caused by the accumulation of mass (or energy). For small objects like a house the effect is so small that we cannot see it or easily measure it.

By now you will not be surprised to hear that strange things happen to time too. Time near a Black Hole (which has extensive mass) passes much more slowly than time on a rocket orbiting it far away. This was clearly illustrated in the film, Interstellar.

3. How do we know that Einstein was right?

Whereas Einstein had done all the hard work, before he could take the accolades he had to show which experiments would prove his theory and these had to be executed on an independent basis.

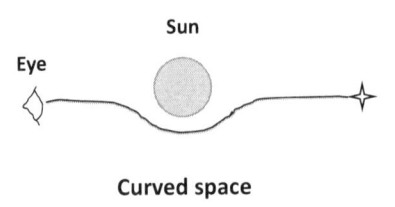

Curved space

Four years after the Theory of General Relativity was published Sir Arthur Eddington's team proved that it was indeed verifiable by experiment. They went to West Africa in 1919 and watched a star disappear behind the eclipsed sun. As shown in the figure above, the star remained visible longer than was expected.

Light from the star was following the curvature of space (and time) around the sun. There are some questions about the accuracy of these early experiments but the principle has been verified many times since.

This new theory also provided new predictions for Mercury's orbit around the Sun. As the closest planet to the Sun the effects of space curvature on Mercury are more pronounced. If this is not taken into account then classical calculations (i.e. using Newton's equations) are inaccurate.

Following the Eddington disclosure, Einstein rightly went on to become a global celebrity and arguably the most famous Scientist of all time.

4. How fast is Gravity?

When a large body changes its mass the warping of Space-time will change and we will perceive a change in Gravity. Newton suggested that this change could be instantaneous. More recently it has been suggested that if it was instantaneous we would experience every change in our universe and living on Earth might be like riding on a boat in a storm.

Einstein said that this change cannot be instantaneous and he reminded us that we have a Universal Speed Limit which must also be the Speed of Gravity.

5. Why are we so sure about the speed of Gravity?

Scientists have recently measured gravitational effects within the solar system to confirm that gravity does move at the Universal Speed Limit.

In 2003 (Jan 7), New Scientist magazine reported that Ed Formalont and Sergei Kopeikin had shown that Gravity moves at the Speed of light. Radio waves from a distant star (actually a

quasar) were bent around Jupiter (in the same way as light bent around the Sun in the Eddington verification for General Relativity) such that the time taken for the star to change position could be measured.

Yet again Einstein's Theories were shown to be correct.

Above we asked 'How long would it take in our Solar System [for space to settle down if the Sun suddenly disappeared]?' Well, if the Sun disappeared then it would take 8 minutes for us to know this (assuming the Universal speed Limit across 150 million km) and then we would experience a roller coaster ride (like the bowling ball on the trampoline) whilst warped Space-time settled down into a new steady state.

NB. Without any Sun this would be the least of our problems!

6.   But is Gravity Real or Perceived?

We now know that Gravity is the force perceived due to the existence of Space-time:

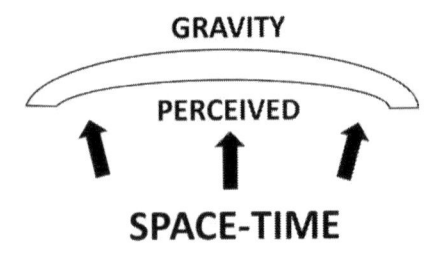

GRAVITY

PERCEIVED

SPACE-TIME

REAL

It is Space-time that creates curvature which gives the illusion of the force of Gravity. As such Gravity is just a property of Space-time (it has no meaning outside of Space-time).

7. Where does that leave Space and Time?

So far we have been thinking about space and time separately but, in his 'Theory of General Relativity' Einstein shows that we should consider them together as one entity called Space-time.

a. Firstly, Einstein showed us that Space (alone) is not absolute (it is different for different people)
b. Then he showed us that Time (alone) is not absolute (different for different people)
c. However, with the 'Theory of General Relativity' Einstein brings everything full circle and gives us something that is absolute (i.e. the same for everybody); it is **Space-time**
d. It is **Space-time** that is Real giving the perception that Space and Time can exist on their own (but they cannot)

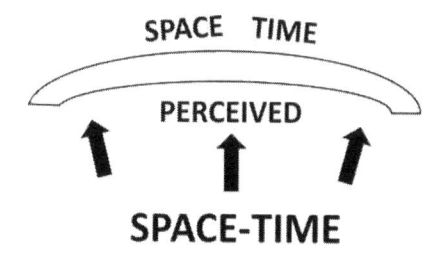

SPACE   TIME

PERCEIVED

SPACE-TIME

REAL

8. How fast is Space-time?

If you are standing still, how fast is time moving? Now if you start walking away from that spot is time moving at the same speed? We can now answer those questions, but you will need to free your mind of any preconceptions:

a. Gravity moves at the Universal Speed Limit
b. But Gravity is not Real; it is something that we perceive due to the curvature of Space-time

c. It is **Space-time** that is Real
d. Therefore it is Space-time itself (not Gravity) that is moving at the Universal Speed Limit (i.e. the Speed of Light)

We know that a photon travels at the Universal Speed Limit through Space-time. Therefore, given what we know about relativity, from the perspective of a stationary photon, Space-time is flying by the photon at the Universal Speed Limit.

I know that it is not possible for a photon to stand still but this is a Thought Experiment and we can push the boundaries.

9. **What is the relationship between Space and Time within Space-time?**

The idea below was first tabled by Brian Greene whom you will find included in my Further Reading list. The relationship based upon speed through space and time is displayed in the graph below:

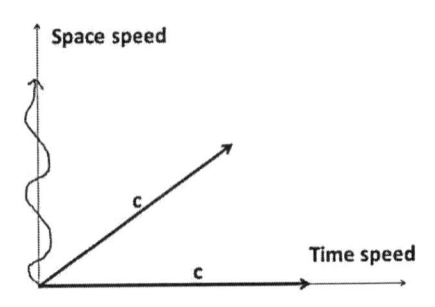

Space-time Speed Graph

a. This is not a Distance v Time graph which we use to calculate speed. It is a directional graph for Space-time Speed which is a graph of Space speed v Time speed
b. On the graph space speed moves on the vertical axis and time speed moves on the horizontal axis

c. If you can stand still (i.e. no movement through space so you are stuck on the bottom axis) - Time will move at the speed of light (depicted by the thick line at the bottom marked with a speed of 'c')
d. If you can move in zero time (i.e. stuck on the vertical axis) then you will be shown moving straight upwards on the graph (this is what a photon does - depicted here as a wavy line)

10. What happens if you move through Space and Time together?

This is what we do in our lives and in the forthcoming chapters we are going to see where we fit on the Space-time Speed Graph.

If you move through space and time with equal speed then your journey is depicted by the diagonal line (as shown on the above diagram). In Notes 4 we take a look at the two extremes on this graph and we will also return to them in chapters 8 and 9.

We know that when one thing changes (Space or Time) something always happens to the other. All of your intuition will tell you that this is nonsense, but it is true and in later chapters we are going to see the proof.

When you start to move, some of the Space-time speed will be transferred to speed through Space and the speed through Time will slow down.

Given this knowledge we can construct a new Law of Nature as follows:

> **Space and Time always move such that together Space-time always travels at the Speed of Light**

It is likely that space and time, as Einstein has taught us to know them, have been unchanged for a long time. If we are to understand what could have been a catalyst for their emergence then we need to know something about the beginning and the evolution of our universe.

6 How did our Universe begin?

When you look at the stars everything looks static. If you watch for long enough you will see stars move across the sky but, when you come back tomorrow night exactly the same thing happens again (a daily light show if you have clear skies where you live).

The night sky that we see is more or less the same night sky seen by every human that ever lived. It is therefore not surprising that, until recently mainstream scientific theory supported the principle of a static universe.

Even though his Theory of General Relativity told him otherwise, Einstein would not go against the science grain on this subject (arguably, he was already in enough trouble). He even added a random factor (later called the **Cosmological Constant**) into his famous 1915 equations to prevent them looking like a new expansion prediction. He later said that 'this was my greatest blunder'.

1. Why did we question a static universe?

In the 1920s Edwin Hubble started to publish data that suggested all was not as it appeared. He was one of the first astronomers to look beyond our galaxy and into deep space where there are many other galaxies which appear to be moving away from each other.

Famously, Fred Hoyle, representing the old school of thought which was grounded upon the concept of a static universe, entered into a public argument with Hubble but he was always struggling to keep up with the new data (mostly showing galaxies moving away from us) appearing from all quarters.

By 1960 Hubble had won this argument but he died (in 1953) before he could be nominated for the Nobel Prize and the Nobel Prize is not awarded posthumously.

It took his namesake satellite telescope in 1997 to really show us what was going on and those involved with that did receive the Nobel Prize in 2011 (Perimutter, Schmidt and Riess). This is explained best by considering an inflating balloon:

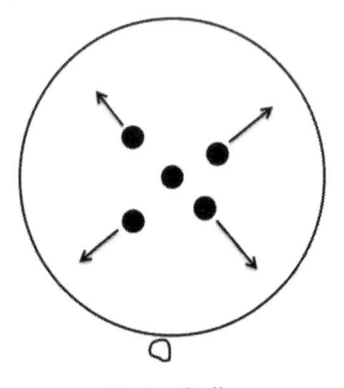

Inflating balloon

Try to ignore what is in the middle of this balloon and just consider the surface. The expansion of the universe is similar to dots on the surface of the balloon. As the balloon is inflated all of the dots move away from each and, most importantly, none can be said to be at the centre.

2. How do we know that our universe is expanding?

Light travels through our universe in waves. To understand more about waves it is useful to look at sound waves. Sound propagates much slower than light so we can detect slight differences with ease.

If we simply stand by a train line or in the line of an approaching aeroplane we can hear it arriving from a long way off; noise from the engine will travel at the speed of sound into our ear.

If the aeroplane is moving towards you faster than the speed of sound then after it sends noise from its engine it moves ahead of it towards you. As it passes you all of these waves arrive in quick succession (creating a **sonic boom).**

Let's now assume that the train or aeroplane is moving at about half the speed of sound. The train will emit a noise and then move closer to you. The train doesn't beat the sound wave but it does squeeze it up a little. This sound is higher pitched.

As the train moves away it stretches the sound waves a little. The change is not appreciable but it is enough to hear a lower pitch.

In the same way that sound waves get squeezed and stretched we observe the same effect with light. Light approaching us is squeezed making its wavelength slightly shorter. This moves it towards the blue end of the visible light spectrum (it is called a **blue-shift)**. It is still travelling at the speed of light because the frequency is adjusted too.

A light source moving away from us stretches the light wave making its wavelength slightly longer. This moves it towards the red end of the visible light spectrum (it is called a **red-shift** and this process is known as the **Doppler Effect)**.

We now know that some events in our universe have consistent similarities. We know that Supernova always explode with the same luminosity. It is the same for rotating neutron stars (aka **pulsars)**. Because we know how bright they should be we can calculate how far away they are. We can also measure any changes to their wavelength (their **Doppler shift)**.

When the Hubble space telescope undertook accurate measurements in 1997 it showed that most galaxies have a red-shift which meant they were moving away from us.

It seems that this movement had slowed but started again about 6 billion years ago. Also, the galaxies furthest away are now moving faster so the conclusion is that our universe must be expanding AND accelerating.

(The latest measurements show that the universe speeds up by 66.9 km / sec for every 3.26 million light years (aka **mega parsecs**) further away.)

3. How did the universe begin?

With the knowledge that the universe is expanding it didn't take long to look at the effect of reversing this evolutionary model. Sure enough, if we rewind far enough back we eventually find a beginning. It was a Catholic priest called George Lemaitre who led these discoveries and it is his name that is associated with the event we now call the Big Bang.

Today the popular scientific view is that the Big Bang created the universe. This event is a location in Time, not Space.

However, we should say that there is a considerable ground swell of opinion in favour of a **Cyclic Universe** and, at some stage this might reclaim the lead in popular opinion from scientists. Roger Penrose is a senior proponent of this view.

Whatever the truth of the underlying fundamentals, everybody agrees that the universe we presently inhabit (I call this our universe) is still expanding from a definitive point in time which we can still call the 'Big Bang' and this was approximately 13.8 billion years ago.

Einstein's famous equation $E=mc^2$ shows that mass (m) can spring into existence from a field of energy (E).

> **This is the most persuasive argument for the 'Big Bang' apparently creating something (i.e. matter) from nothing (i.e. things we cannot see, like energy)**

If God has a place in this book then it is here. God may have been involved at the B of the Big Bang (i.e. right at the beginning). We currently have little idea of what started the Big Bang. Many scientists, however, doubt that God needs to be involved at all and, eventually we will be able to explain everything in scientific terms.

4. So, that is that then, is it?

The Big Bang was something the public could easily understand. You will have known about it before you started this book. However, the scientists that worked on this theory soon found many questions they could not answer.

There were two key questions that badly needed answers; these were:

1) Why does our universe look the same everywhere?
2) Why is our universe the same temperature everywhere?

5. Why does our universe look the same everywhere?

When we say that everything looks the same in every direction this is not literal. Clearly, the Plough constellation is different to Orion but the night sky in every direction looks very similar. There are a similar number of stars and galaxies in every part off the celestial dome above us. This is called the **Flatness** problem.

Clearly, this would not be the case if the universe started at a point and expanded in one direction only. Also, we have no evidence that our position within our universe is in any way special (**Copernicus** won that argument over 450 years ago).

6. Why is our universe the same temperature everywhere?

The second problem was to do with temperature.

The early universe would be very hot and it would appear to exist as a gas. The temperature of a gas is the same as the average speed of its atoms (or molecules). The temperature is changed when a second gas is introduced and it has atoms moving at a different speed. There is a physical interaction.

The molecule collisions eventually have everything moving at about the same speed and this will be the new temperature. Now what happens if the molecules are far apart? What about if they are further than many light years apart (as they are)? How do they communicate with each other if they can't bump into each other?

The expanded universe appears to be the same temperature everywhere, so how did this happen? This is known as the **Horizon** problem.

7. What happened immediately after the Big Bang?

Inflation was a way to answer some of these questions. It states that shortly after the Big Bang there was a rapid expansion during which space <u>itself</u> was created:

a. The figure overleaf (which is not drawn to scale) describes Inflation. It is a distance/time graph because I want to show how remarkable this event must have been
b. A big change to normal distance/time graphs is that the time axis is shown with exponential measurement; the numbers shown are the powers of 10 measured in seconds (see Notes 5 for an explanation of powers).

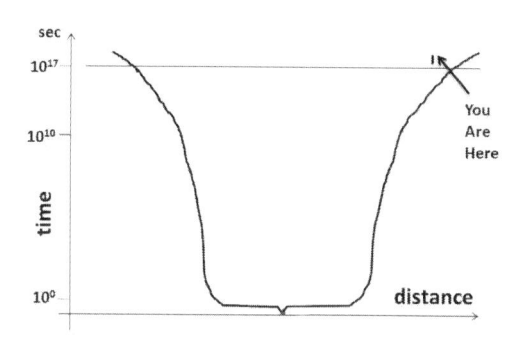

c. 10 to the power 17 seconds is equivalent to about 9 billion years. We are just above that today. The line top right represents the duration of our solar system and we, the human race (aka Homo sapiens) having existed for about 100,000 years, will be a pin prick at the top of the line.

Diagram model of Inflation

d. We use an exponential scale so that we can see all of the time, from 1 second (10^0) to 14 billion years.
e. When we do that we can see how amazing it is that all of the Inflation occurred faster than you can say Big Bang (i.e. less than 1 second).
f. For exponential graphs we expect to see steeply increasing sides (as shown in Notes 6). However, instead of being shaped like a wine glass the graph above is shaped more like a beaker so something strange is happening here

The theory requires expansion faster than the speed of light. If he had been alive Einstein would have been horrified. However, scientists would have told him that nothing was moving within Space faster than light; it is Space itself that is breaking the Universal Speed Limit. So that's alright then (is it?).

Recent estimates suggest Inflation could have started at a pin head with about 10 kilograms of condensed matter – imagine atoms with all of the gaps taken out (remember 99% of the atom 'appears' as empty space).

The rapid expansion of space resulted in the creation of the universe we see around us today. In less than 1 second we get a ready-made universe which is uniform in all directions (as seen with our telescopes). Inflation forecast that radiation from this period of the universe's evolution would be of similar temperature everywhere. This was the prediction which would make or break Inflation Theory.

location

Note that Inflation requires Space <u>itself</u> to be created. This book should also convince you that Time must have been created simultaneously (i.e. Space and Time have no meaning when separated). Ergo, Big Bang Inflation Theory requires that Time had a beginning in our universe. This means that there was a time when there was no Time!

Alan Guth was getting ready to leave academic life when the Theory of Inflation suddenly occurred to him. He quickly changed his mind and published his new theory.

8. What happened after Inflation?

The next phase took about 380,000 years. This is the period during which the fundamental building blocks of matter went crazy. Everything was too hot to become stable and form atoms:

a. This particle soup is also referred to as '**Plasma**' - Protons, Neutrons and Electrons all move around separately in blissful chaos
b. As our universe cooled this crazy motion terminated in the formation of the first permanent atoms
c. Note that the density of matter within our universe reduced rapidly and that also had a shock effect on everything that was around at the time
d. Electromagnetic radiation started at the end of the Plasma period when things started to settle down and atoms became stable – the beginning of what I call the 'Light Age'

When this radiation was discovered it provided the evidence to substantiate Inflation theory. Penzias and Wilson stumbled upon this radiation whilst looking for something else. They thought their telescope was inaccurate and scrubbed bird poo off it because they thought they were receiving unwanted 'noise'. Instead, it turned out that they had found what everybody else was looking for. Being lucky is a common way to win the Nobel Prize.

What they found was Cosmic Microwave Background radiation (CMB). It was hiding in the Microwave section of the Electromagnetic Spectrum. It was about $-270°C$ and very uniform. Although you would not know from the exaggerated pictures we see of it (see figure overleaf) CMB differs by less than one thousandth of a degree centigrade across the entire sky.

Cosmic Microwave Background

CMB is believed to be a snapshot of what the universe looked like at the beginning of the 'Light Age'. Moreover, because of the immense speed of Inflation it is also effectively a message of what was in place at the Big Bang. Consequently, there is a considerable amount of analysis of the latest CMB images which were taken by satellite away from any interference in our atmosphere.

The cataclysmic events of the early universe were over but there was still some evolution required to get to our present situation.

7 How did we get Here from There?

After the inception of the 'Light Age' we can see what happened next. New atoms from the Plasma period started to congregate into local areas. If you have a slate floor and put off cleaning long enough you will see this effect with dust. It naturally forms lumps. Experiments with mud display the same tendencies and it is gravity in action on a tiny scale.

These local areas became our early galaxies and within them stars started to form. Our telescopes can detect new galaxies that are forming in this way and, if you haven't got a telescope, you can see amazing photographs of new galaxies forming (aka '**Nebulae'**) like that below:

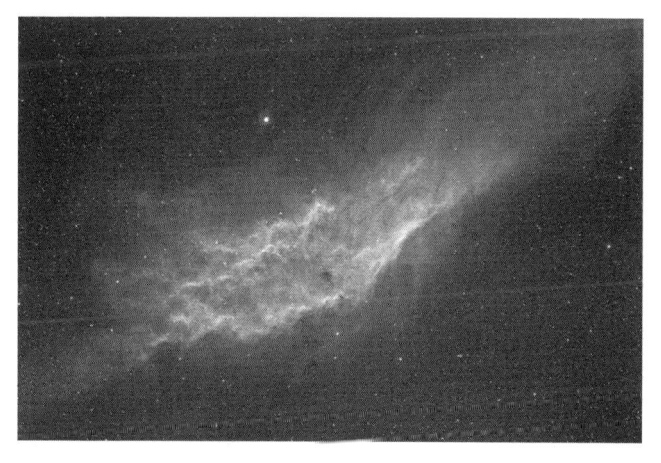

California Nebula

We are now aware that there are about 2000 billion galaxies and that each of them has, on average, about 400 billion stars. Are those numbers significant? Not really. Nobody has ever counted them all but, we will, over time get ever more accurate figures.

The significance is that they are both (i.e. the number of galaxies and stars) very big but not infinite.

Galaxies come in all kinds of shapes and sizes. Ours is a spiral galaxy called the Milky Way. It is shaped like a saucer and about 120 thousand Light Years across with most of the stars and all their planets on the same plane. We have never seen our galaxy from far away but we have seen many other galaxies this shape and all of the data suggests that we are living in a similar one.

Voyager 1 was redirected to leave our galaxy at about 90° to the plane of the orbits of the planets so it is on course to look down on us and our galaxy. It is the furthest manmade object from Earth (about 142 times the distance of Earth from the Sun in 2018) but it is unlikely that it will stay awake long enough to show us the picture we all want to see – the Solar System from above.

1. What do we know about our Solar System?

Our Sun and its planets cooled and solidified into its current profile about 4.5 billion years ago. If you are interested in how the Sun works see Notes 7.

Our Solar System is situated on a minor spiral of the Milky Way galaxy. From here we can see that there are a lot of stars clustered together in a central area. People call this the Milky Way which is not accurate. All of the stars we can see with our eyes are part of the Milky Way and these are everywhere we look. It is true, however that there are more near the centre than elsewhere and that is why we see a 'milky strip'.

Medium sized stars (like our Sun) live for about 10 billion years so, at 4.5 billion years old, our Sun is middle aged. When all of the hydrogen is used up (through **fusion**) our Sun will expand into a Red Giant which will engulf Mercury, Venus and the Earth. When all the helium is used up it will collapse into a White Dwarf and then fade into a Black Dwarf.

Life of a Medium Star

2. What about other stars?

Small stars, which have low mass, remain relatively inactive for tens of billions of years and then, when their hydrogen is used up, they also collapse into a White Dwarf.

Large stars are anything bigger than about 20 times our Sun. These are the furnaces in the universe which make all of the elements we know about. A complex process of squeezing atoms together results in a stock of every element up to iron.

The star then explodes as a Supernova which creates all the heavier elements we know about. The outer layers are blown into the cosmos and will add to the dust clouds which will coalesce into a star system like ours. These events are quite common within the universe (and are used to measure its expansion) but they are too far away to spoil our enjoyment of life.

Large stars have a relatively short life (measured in millions of years) which is just as well because we needed all the elements they gave us to create life. Some of the elements that are essential for human life (such as Carbon and Oxygen) were made in this way.

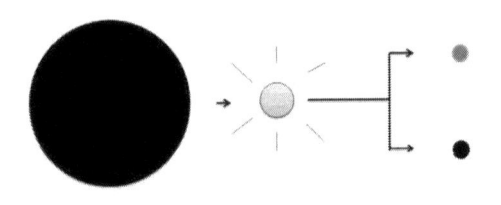

Life of a Large Star

Hydrogen was the first element to appear from the early Plasma period in the universe and it has got everywhere ever since. The rest of the elements that blew in to our Solar System were made in large stars.

Therefore, given that we are mostly made from carbon (and other elements that like to be near it) we are mostly made of Stardust (in fact about 93%).

3. All roads lead to a Multiverse

We know that we are privileged to be here in our Solar System but our route towards life raises many questions. In my book, 'Physics Mysteries for Absolute Beginners', I analyse the Fine Tuning problem; is it a coincidence that so many factors are just right for life?

The Multiverse argument states that all the possible Fine Tuning that can happen, does happen. Consequently, there are many different kinds of universe. Perhaps something happens at a Big Bang to determine what the parameters will be.

Clearly, we live in a universe that supports carbon-based animals because we are carbon-based animals. If we were lithium-based animals we would need to live in a universe that supported the easy manufacture of lithium, but we do not.

We should take note that, because of a lack of evidence, some scientists (including Sabine Hossenfelder) believe that the Multiverse hypothesis is as absurd as the Fine-Tuning problem it tries to explain.

4. What is a Multiverse?

A Multiverse is simply two or more universes. A number of topics, covered in this book, are still giving scientists problems they cannot solve.

The Multiverse is a new Theory that appears to provide the additional explanation required. Many scientists now believe that, if Inflation could provide the foundation for our universe then it is likely that it would do the same for others.

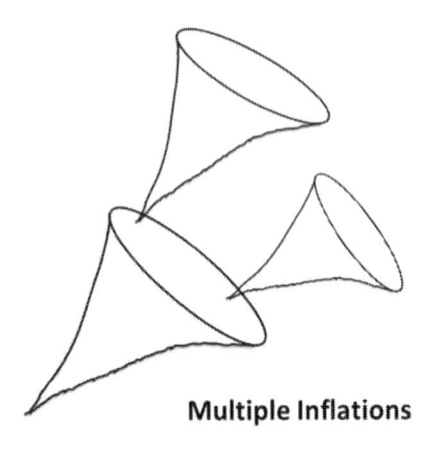

Multiple Inflations

Note that these 'Multiple Inflations' are cone shaped as opposed to the Beaker (flat bottomed) shape of Big Bang Inflationary Theory shown earlier. This is because this newer theory does not need to worry about what happened before the Big Bang. Everything is continuous and we don't need Big Bang Inflation (which has things moving faster than the Universal Speed Limit).

Each new act of Inflation effectively has a Big Bang but, given the continuous progression involved, scientists refer to them as **Big Bounces**.

We have figured out how our universe has managed to evolve to its present state. We may need to tweak the theory a little but, in general, it has been a magnificent feat of intellectual ingenuity.

So we have some reasonable theories and some proof. The main questions now outstanding are what caused events to happen in this way; what was the catalyst for the Space-time that we experience?

8 What was the Space-time Catalyst?

In this chapter we look at how major changes could happen. We have already seen that our universe must have progressed through various stages to get to where we are today but what causes it to make those changes. This remains a scientific mystery and that is why we have used it as the central theme within this book. This chapter will also give us a view as to how scientists start to confront these big questions.

1. **What are the significant change points within the universe?**

Firstly, we need to understand what is meant by Symmetry:

a. You will probably be familiar with lines of symmetry; the figures below show 1 line of symmetry for the arrow, 4 lines for the square and an infinite number of lines for the circle:

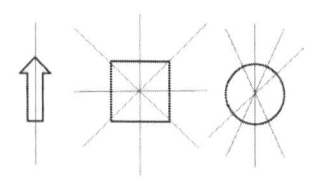

Lines of Symmetry

b. We can have different sorts of symmetry (i.e. rotation, inversion etc.) but we will just focus upon the simplest which is positional (or **translation**)
c. Positional symmetry compares what we find at one position with another - for example if our experiment works on the

kitchen table it should work on the dining room table too (principle of **invariance**)
d. In general, if we move anywhere within space (i.e. up, down, forward, backwards, left or right) we find that the Laws remain intact - everything stays the same
e. We are very used to this situation but we have already identified three occasions during the evolution of the universe when this was not the case. These were:
   - Big Bang
   - Inflation
   - Plasma Period
f. These all mark an epoch where there was major change and we now know that the Laws of Physics can be created (or changed) when there is a major change within our universe
g. Scientists call this a **Break in Symmetry**

The Theory of Special Relativity was an analysis of the Symmetry of Constant Speed (aka **Boost Symmetry**). This means that as well as being able to duplicate the experiment on your dining room table you can also duplicate it on a table in a train moving at constant speed.

2. How much of this applies to fields?

What exists within our universe? The answer is Matter, Light and Space-time (it's a short list). They all have one thing in common which is maybe not obvious from where you are sitting:

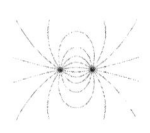

a. Space-time is the warped structure of the universe and, as you can see from Chapter 5, it is a **field**
b. Light (or electromagnetic radiation to more precise) consists of electric and magnetic **fields** and this explains its propagation (see Chapter 2)

c. We have also intimated that matter particles can also be regarded as **fields**; you will need to read 'Quantum Physics for Absolute Beginners' to confirm this
d. So, our universe is just something full of fields and these fields all have something in common – Energy

3. Was cooling a Catalyst?

The first field event occurred at the Big Bang and we can be pretty sure that it was very, very, very hot. Consequently, arguably the next thing to happen in our universe as it got bigger was the process of cooling.

When the early universe was millions of degrees hotter there will have been situations that we have never seen. Scientists are still trying to understand how 'plasma' functions.

The early universe would have been a huge game of tetris; imagine the building blocks of matter (we have covered electrons, protons and neutrons in this book), join them together in any way that you want and that was probably how it all started.

Scientists also believe that a super cooling process (such as what probably occurred during Big Bang Inflation) may have had further Symmetry Breaking effects.

4. What about Density reduction?

a. At the same time that the universe was cooling it was expanding and also becoming less dense
b. Density has a direct connection to Space-time because it is mass that gives the Space-time field curvature
c. The density of mass-energy within our universe is measured by a constant called omega (Ω)

d. If omega is too big (much bigger than 1) our universe would contract under the influence of Gravity
e. If omega is too small (much smaller than 1) the universe expansion would be too quick for galaxies to form

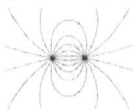

f. Scientists have measured omega to be about 1 (i.e. not too big and not too small but just right)
g. However, this may not have been the case for ever. Perhaps a sharp density reduction was the Catalyst for the emergence of Space-time (as we know it)

5. **How many dimensions are there?**

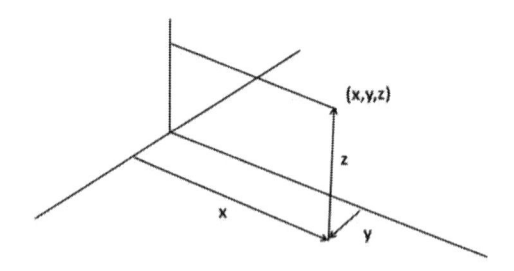

Three Spatial Dimensions

You may be familiar with the common way of displaying three dimensions all at right angles to each other (see the figure above) – we call it a graph and all axes intersect such that:

a. If you move x way down the x axis, y way down the y axis and z way down the z axis we reach a point which we uniquely identify as (x,y,z); this represents the three spatial dimensions that describe the universe we see

b. If we place a time stamp (t) at (x,y,z) then we have the four dimensions that we use to describe everything around us (x,y,z,t)

In thought experiments you do not have to constrain yourself to four dimensions. If we describe a point as (a,b,c,d,e,f,g,h,i,j) we have now included ten spatial dimensions. Luckily, mathematicians are ahead of us and they have been performing calculations with points like this for some time. So, adding extra dimensions to support a theory is not a problem.

All you are doing is stating that something (a point) needs ten pieces of information to locate it. There are some scientists who believe that there were multiple dimensions during the Plasma period and that, following a Break in Symmetry, the universe collapsed into the four we see today.

The Perimeter Institute in Toronto have postulated that charge itself might be a dimension left over from the Plasma period. As theories go this is right out there but, at least, it tries to explain something about charge which is one of the more mysterious manifestations in nature (I think).

6. Why do we only see three spatial dimensions?

Take a look at the figure below:

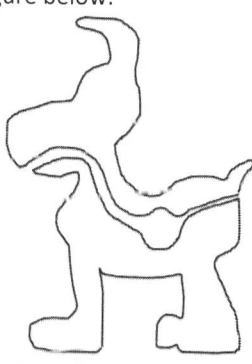

Two dimensional dog

If we only had two dimensions, clearly dogs (and humans) would fall apart. We need three just to keep our bodies together. So, how did we end up with three spatial dimensions?

Perhaps there were more at the outset and some kind of condensing activity removed all of the other dimensions from sight, leaving just the three we see in our macro world today.

Of course, the other dimensions might be hiding somewhere? This is the third item that could have been a Catalyst so the list is:
- Cooling through expansion
- Density decrease
- Dimension reduction

7. Where does the universe go from here?

Firstly we should say that what we see right now will continue for a very long time; at least for as long as the extent of our species. However, in a cosmic time period things are going to change. Everything is going to slow down and get even cooler. Stars will eventually disintegrate.

The Law of Disorder states that the disorder in the universe will increase. When we reach a maximum position of disorder everything will be at the same speed and temperature. Scientists will tell you that if you are unable to create a temperature difference you cannot enact any process.

When the universe reaches this position of equilibrium we will have reached '**Heat Death**'. This does not mean no temperature but rather there will be no temperature difference anywhere, which means no activity.

8. What about the Cyclic Universe Theory?

If we ignore the scale of things (i.e. the eventual universe versus the tiny beginning), this 'Heat Death' has a lot in common with many characteristics of the Big Bang:

a. The cooling and emptying, thinned out universe will be mostly reduced to photons and, as we know, photons do not recognise a passage of time (as we do)
b. The universe may appear to be huge but, remember we said that the Big Bang (i.e. the start of what we experience today) was an event which was a location in Time, not Space; if there was a Break in Symmetry which allowed all the remaining particles to effectively disappear (or lose their mass, at least) then we might have the situation where we have Space but reduced Time
c. We have already seen something like this when we introduced the Space-time Speed Graph:

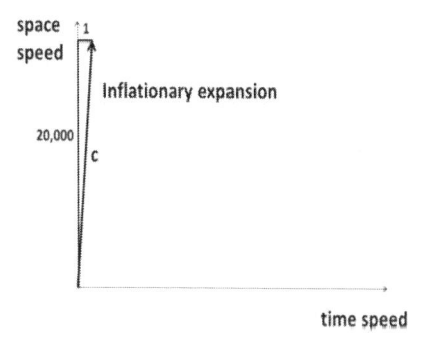

```
space  ↑1
speed  │↑
       ││  Inflationary expansion
       ││
20,000 ││
       ││ c
       ││
       ││
       ││
       ││
       └└─────────────────────→
                          time speed
```

SPACE-time Speed Graph

This is the scenario on the Space-time Graph where Space speed is dominant (we refer to it as SPACE-time in Notes 4). Our graph considered a ratio of 20,000:1 in favour of space speed but, of course, the bouncing universe could have been even more extreme.

We have spent most of this book showing that Space or Time cannot exist on their own in our universe. If an event occurred where Time was removed maybe a new Break in Symmetry would recreate a new Space-time (where Time re-emerges) and there would be a new era or a new universe.

Alternatively, it is reasonable to consider that Time in a Cyclic Universe might not have a beginning but that instead it is infinite (i.e. during the transition Time is reduced to an infinitesimal amount but not removed). A growing number of scientists prefer this possibility to the alternative option where Time has a beginning.

Note also that this process does not require anything to travel faster than the Universal Speed Limit which means that Einstein can rest in peace.

At this point a Cyclic Universe could set new starting conditions. We will obviously live in a universe that supports carbon-based life. Perhaps all of these universes support carbon-based life but then again, perhaps you have to be n universes in (where n is a number).

You now have most of the information that you will need to address the challenges we set in Chapter 1:

- **If Space and Time are Perceived then what is Real?**
- **What changed to make it Real?**
- **When did this happen?**

Let us now make sure we really understand Space-time.

9   What is Space-time?

Firstly, Einstein gave us the 'Theory of Special Relativity' which showed us that we had to relearn everything we knew about Space and Time. Ten years later he released his greatest work, the 'Theory of General Relativity' which included Gravity. The central insight within this masterpiece is that Space and Time are part of a double act, called Space-time.

This chapter will take a step back to make sure that you now understand Space-time. Watch out for the trick question at the end! So….

1. What has Einstein taught us, again?

a. In his Theory of General Relativity Einstein tells us that space is curved around mass and that large masses like our Sun make this visible
b. This curvature acts upon other objects to determine their path through space
c. The amazing implication here is that Gravity is not a special force but just the result of a Space-time field with an intensity gradient caused by the accumulation of mass
d. Space and Time cannot exist on their own
e. We now know that we cannot think about Space and Time separately but that we should consider them together as one entity called Space-time; it means that when one thing changes (Space or Time) something always happens to the other
f. As Space becomes curved there will be an equivalent intensity gradient effect to Time (which will slow down as we get closer to a large mass)
g. Also, Space and Time are not the same for different observers (they are not **absolute**)

2. Is Space-time Curved?

Yes, when it is under the influence of mass or energy. It is this curvature which creates the force of Gravity that we Perceive.

3. What is the curvature of Space-time away from large masses?

a. This is a good question
b. Firstly, remember that Space-time is a field
c. The shape of this field can have three possible answers:
   - It could be curved like the surface of a sphere (i.e. convex)
   - It could be curved upside down like the surface of a saddle (i.e. concave)
   - It could be flat
d. As far as we know Space, under no other influence, is flat
e. This is very convenient; any other answer would make calculations extremely difficult

4. How big is Space-time?

A good way to answer this question is to look at the space contained within our Galaxy. Because the universe is so vast we use a Light Year (the distance that light travels in one year) as a measurement of distance (it equates to about 9.5 x 1000,000,000,000 km). What would be required to travel to our nearest neighbour?

Our nearest neighbour is a star called Proxima Centauri (part of the Alpha Centauri group) which is about 4 Light Years away.

10 years after being launched, the exploration rocket New Horizons flew by Pluto in 2015 at about 14 km/sec. This is the fastest man made object built so far and, relative to other vehicles on earth, it is very fast - you would cross the Atlantic Ocean in 5 minutes.

It may seem fast to us but New Horizons is about 20,000 times slower than light (which travels at 300,000 km/sec). Therefore travelling to Proxima Centauri using current technologies will take about 80,000 years (i.e. 4 Light Years x 20,000).

Clearly, we will need a 'Star Trek' like break through if we are to explore anywhere outside of our solar system whilst our species is still alive. Our galaxy is about 120,000 Light Years across. Our universe is vast and even our nearest (star) neighbour is out of reach.

5. Can we travel through Time?

a. Ah, the Big Question
b. Yes we can – we are all doing it right now (no? not that Question)
c. Which Question then - Can we travel into the future?
d. Yes we can – because we can slow time down by going faster or experiencing greater Gravity, then, if we do the reverse (i.e. go slower or move away from Gravity) we will speed time up and therefore move into what you would regard as the future. Neither of these things is easy though!
e. Still the wrong Question? Yes. What we really want to know is can we travel back to the past and relive something that appears to have already happened?
f. Ah, now it's getting tricky. Everything you have learnt suggests that this should be possible (and the Laws of Physics do not prevent it) but ….
g. We get the paradox problems. The most obvious one is could we return and kill our father before he meets our mother? If we could, how could we be here to witness it?
h. There is also a problem with disorder. If we go back and make something more disordered (like scrambling an egg) we don't expect to find it ordered (still intact) when we return to the future

i. This and other similar paradoxes suggest that we cannot travel back through time and, even if we could, there appear to be natural laws (that have not been codified yet) that prevent these paradoxes – otherwise we would see them around us every day and we don't

6. How fast is Space-time?

We have already seen (from Chapter 4) that Time cannot be measured. We can, however, compare different timing mechanisms (i.e. pendulum and quartz clock). We also know that Time is a personal commodity (i.e. not **absolute**) so it is not really surprising that we have no generally agreed Time measurement.

However, this is not the case for Space-time and in Chapter 5 we showed that we have a clear answer to this question:

a. Space-time moves at the speed of light - what does that mean?
b. Given that we live in four dimensions imagine that all the space dimensions (x,y,z) do not move. Imagine that you can stand still. Now time alone (t) is moving at the speed of light
c. When you start to move, some of the Space-time speed will be transferred to speed through Space and the speed through Time will slow down. When Space and Time are apparently separated they move at different speeds
d. But Space and Time, together, always travel at the same speed which is the Universal Speed Limit
e. Note that the only other things that move at the Universal Speed Limit behave like waves with wavelength and frequency; this thought might help later

7. Can we locate our position within Space-time?

Let's get the Space-time Speed Graph back:

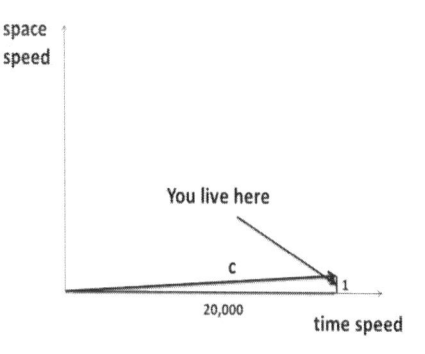

Space-TIME Speed Graph

Nobody knows what the space speed and time speed relationship is within our Solar System. However, we can analyse the extreme positions (as we did with the Inflationary Space-time environment).

At a Black Hole (discussed further in 'Physics Mysteries for Absolute Beginners') scientists believe that Space is dominant and that Time is reduced to a background effect. If that is the case then we appear to live in a place which has the opposite metrics.

The triangle at the bottom of the graph (also described in Notes 4) displays time speed 20,000 times faster than space speed. This ratio was chosen because this is the space speed that equals the fastest man made vehicle ever achieved within our Solar System (i.e. New Horizon which flew past Pluto at 14 km / sec).

This implies that we all currently live on the edge of this triangle (with a combined speed of c) and this is how we perceive Space-TIME. It appears that our world is dominated by Time speed.

8. **Does this make Space Speed and Time Speed another double act?**

   a. Well, let's work it out
   b. Can they exist separately – No, so they are like Matter and Light
   c. Can one be converted into the other – Yes (that's what is happening on the Space-time Speed Graph) so they are like Energy and Mass
   d. Does one induce the other – Probably, so it's like Magnetic and Electric fields too
   e. I think that makes it a double act

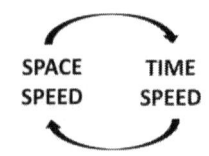

```
     SPACE    TIME
     SPEED    SPEED
```

9. **What is the shortest distance and longest time within Space-time?**

We have now completed our learning journey through Space-time but we still have this 'tricky' question to answer.

Assuming that natural Space-time is flat (see item 3 above) what is the shortest distance from one point to another? Well, a rocket moving at constant speed will cover the minimum amount of distance from one point in Space to another in a straight line (see the diagram overleaf).

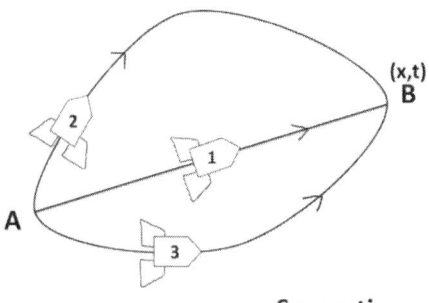

Space-time

Rocket 1 is moving at constant speed on a straight line course from A to B (as shown in the middle of the diagram above) and this is the shortest Space-time distance.

Don't be fooled into believing the diagram is a distance (space) v time graph. If you have to visualise axes then you must perceive them both as Space-time. The diagram is there only to illustrate the conundrum that we will unearth.

Rockets 2 and 3 start (with the same constant speed) at A but then take a different journey (through Space-time).

Which rocket will take the longest time from A to B?

a. If you want you can imagine that Rocket 2 accelerates up to the Speed of Light (or as close as it can get) and then turns around and returns to B. You should not be too surprised when I remind you about time dilation and tell you that Rocket 2 will get to B (with time location t) faster than Rocket 1 (because its time duration will be less)
b. If you want you can imagine that Rocket 3 slows down. However, if they are going to get to B (the (x,t) location in Space-time) then somebody had better wake the captain and

make him put his foot down - if he makes it to B he will also get there before Rocket 1

c. If Rocket 3 misses the (x,t) location then the captain will have lost his chance; he can never visit that time (t) location again (unless he can travel backwards in time)

So far you have been dealing with this problem using Newtonian Physics (aka **Classical Physics**) and his Laws of motion (see Notes 1) but Newton knew nothing about Space-time. You need to clear your head of that old stuff and consider the new stuff. In this book you have learnt a new Law of Physics which is:

Space and Time always move such that together Space-time always travels at the Speed of Light

---

a. Given that we know that Space-time moves at the Speed of Light, if we are using up more Space-time speed through space it means there is less Space-time speed utilised through time (i.e. less time passes)
b. Consequently, given that Rockets 2 and 3 are using more Space-time Space through Space-time they must be using less Space-time Time!
c. Rockets 2 and 3 are going to get to B quicker

In fact, any rocket taking a different route to B will get there quicker. The location in time will be the same but the duration through time will be less. These time definitions are different and Time really is experienced differently by different people.

The reason we do not notice our time differences is that in our world they are so tiny. However, using this thought experiment we can now answer the question raised above; which rocket will take the longest time from A to B?

> **Rocket 1 moving at constant speed will cover the shortest distance from A to B but it will also be in the longest time possible.**

If you don't find this amazing then go back and read this section again. This is the toughest part of this book but it is the world and the universe that we live in. This is how Space and Time complement one another.

Although we have covered this topic in Chapter 5 it is worth repeating it, albeit taking a different perspective. If you are going to find a new Space-time question then you are going to have to reprogram the neurons in your brain and repetitive stimulus will help!

We have now come a long way, from the definition of matter and light up to the speed of Space-time, and I am aware that I have not given you any tangible proof of the Space-time assertions that I have made. We will therefore take a brief break to rectify that situation.

10 How does Space-time work?

If you have been reading this book respecting my explanations and opinions but reserving your right to make up your own mind (at the end) then this will be the defining chapter for you. You can question science theory (and I have encouraged you to do so) but, if you do not accept these new concepts, you will now also need to question the technology that is part of your life.

This chapter focuses upon the latest discoveries and some of our modern technology which is dependent upon Space-time.

1. How does GPS work?

The Global Positioning System (GPS) is a network of satellites orbiting the Earth at 20,000 km all moving at circa 14,000 km/hr and circling the globe twice a day. Although satellites are always being swapped in and out of the network, at any one time there are always 24 available and a user of the system should be able to detect about 12.

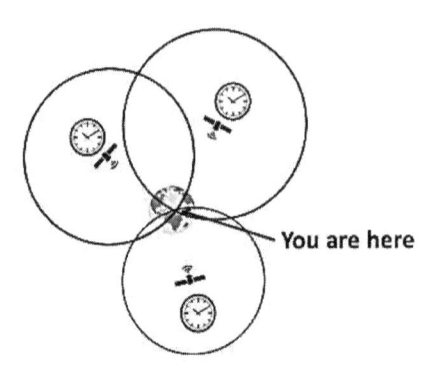

GPS Trilateration

Because GPS will give you a location in three dimensional space (latitude, longitude and altitude) it can be used by pilots in airplanes. Alternatively, you can use it in a Sat Nav system in your car on Earth which is shown in the diagram.

Your Sat Nav receives messages from at least three satellites. Depending upon your location you will often locate more. Each satellite sends an ID, the time it sent the message and its exact location when it sent the message. Because the messages are electromagnetic radiation (radio waves actually) moving at the speed of light, if we know how long the messages have taken we can quickly calculate the distance to each satellite (a method called **trilateration)**. Because your Sat Nav knows exactly where each satellite is and it knows how far it is from each one it can calculate its position to within 3 metres.

Note that this process relies upon exact timings and, if you have learnt one thing so far it is that time is different in different circumstances. For trilateration we have two problems:

a. Firstly, compared to our Sat Nav on Earth the satellites are orbiting in lower gravity. The time at the satellites is 45 millionths of a second per day faster than those on Earth
b. Secondly, compared to our position on Earth the satellites are moving much faster. Their time is slower by 7 millionths of a second per day (see **time dilation**)

After we have taken account of both effects the overall time difference is 38 millionths of a second each day. If we did not take account of this in our distance calculations we would accumulate an error of about 10 km per day and the GPS system would be useless.

(FYI: You may be wondering where these two 'time' effects cancel each other out. It is at an orbit which is about 1.5 times the radius of the Earth.)

2. How does Gravitational Lensing work?

An optical lens receives light from a distant source (i.e. parallel lines of photons) and bends them to a focal point. If we put our eye in this position we get a clear picture (albeit upside down) of the distant object (see **a** in the figure below). Gravitational lensing works in the same way but instead of an optical lens we utilise a large area of mass like a distant galaxy (see **b** in the figure below):

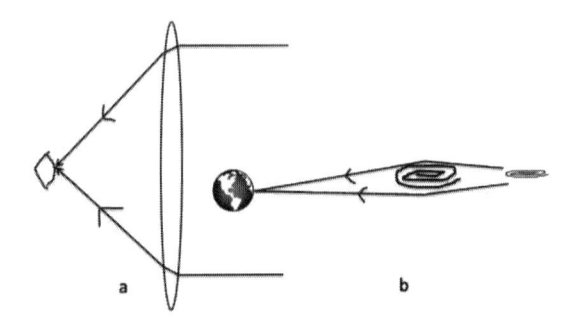

Gravitational Lensing

The galaxy bends Space-time so that light from a galaxy even further away, which we could not see otherwise, gets focused upon our planet. If we are in the right focal line then we will see a ring of light that has been warped and we can then discover something about the galaxy that was hidden.

This type of astronomy is conducted at a number of observatories including Mount St John in New Zealand. It was also a project on Zooniverse recently where the public were asked to identify new galaxies using this technique.

3. Is the search for Gravitational Waves over?

Yes. The definitive proof of Einstein's Theories came in 2016 when the detection of Gravitational Waves was confirmed. To appreciate the work done in this area we firstly need to understand interference. Interference is the phenomenon that allows us to measure distances that are smaller than an atom:

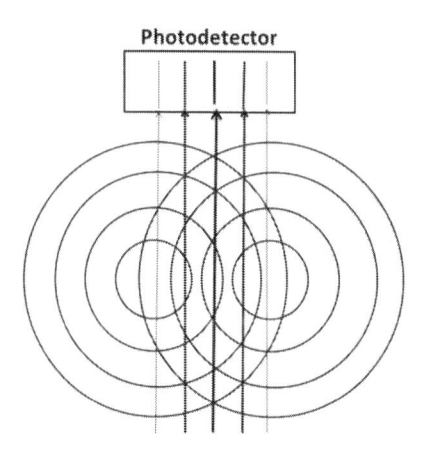

Interference

The diagram above represents two light waves. Each wave will have a peak and a trough but our diagram just shows the peaks. Where peaks overlap they form a peak which is twice the size. These have been joined together on the diagram to show that, as the waves move upwards, these results in a series of straight lines with the most prominent in the centre.

Although not shown on the diagram the troughs will also overlap to form troughs twice as deep and all peaks and troughs that meet will cancel each other out. The final picture, captured by a photodetector, is a series of lines as shown above.

The distance between each peak is the wavelength of a photon, which for visible light is tiny. As you can see in the diagram above the distance between the lines on the photodetector is also about the size of the wavelength of a photon and it is this phenomena that allows us to carry out experiments with high sensitivity in this range. This feature of nature's waves is exploited by an Interferometer.

The Interferometer was invented by Albert Michelson, who in 1887 with Edward Morley showed that the Speed of Light is constant. This definitive experiment, at the end of many decades when most physicists thought there must be a medium which would make light travel at different speeds, was the incentive for Einstein to develop the Theory of Special Relativity. Michelson received the Nobel Prize for his invention of precision instruments in 1907.

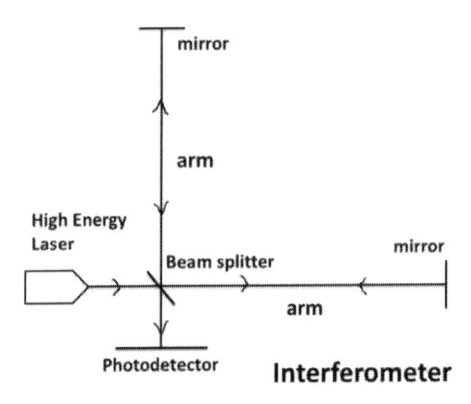

Interferometer

As shown in the diagram above, an interferometer has a light source which immediately splits the beam so that two beams travel through space and recombine through another splitter before being presented to a photodetector which is sensitive enough to read an interference pattern (i.e. the distance of the wavelength for the light source used).

The arms may be adjusted so that they are exactly the same length and therefore any difference in the photodetector interference pattern (or lack of it) must be due to some intervention along the way. The intervention that Michelson / Morley searched for was a medium for light propagation (aka the **aether).** However, all of their results were the same and because they found no evidence for the aether the implication was that the Speed of Light must be constant.

Most recently work in this area has taken place in the USA at the Laser Interferometer Gravitational-Wave Observatory, known as LIGO. Instead of the aether they were searching for Gravitational Waves.

Whereas both experiments use L shaped arms with beam splitters and mirrors just about everything else is different. The arms at LIGO are 4 km long (Michelson had 11m arms), it uses high energy lasers which means that wave interference is sensitive enough to measure a movement which is one ten thousandth of the size of a proton.

All scientific experiments gain in credibility when they are replicated. For LIGO replication is built into the process. LIGO has a sister experiment 4500 km away which is required to show the same results if an event is to be reported.

Most importantly, LIGO was not measuring a feature of nature on our planet (such as the Speed of Light); it was waiting for an epic event (such as Black Holes in collision) to occur within our universe so that we can see the effects that this has upon Space-time everywhere.

The best analogy I have for Gravitational Waves is an Earth based Tsunami. After a significant disturbance (such as an under-water eruption) the Tsunami travels across the ocean until it collapses on land. If we have buoys waiting in the ocean we can measure the size and speed of the Tsunami Wave as it passes.

In a similar way, a significant disturbance in the universe will create a Wave (Gravitational this time) and it will continue throughout the universe until something stops it. Along the way (like the buoys) LIGO is able to measure the size and properties of a Gravitational Wave.

LIGO was expected to spend most of its time waiting. This ambitious experiment started in 2007 but for eight years it showed no signs of Gravitational Waves. However, in 2015 an Advanced LIGO, which includes improved techniques for handling seismic disturbances, allowed the experiment to increase the search area to 225 million light years and, as soon as this was turned on, they got lucky.

Finally, in February 2016 the science community was delighted to hear that LIGO had detected Gravitational Waves.

Scientists also calculated the shape of gravitational waves that were expected to permeate the universe. The Gravitational Wave pattern detected also proved that Black Holes exist, something few doubted but could not explicitly confirm.

It is now believed that LIGO and other Gravitational Wave research centres, which can focus upon different wavelengths, will be able to analyse other scientific questions such as:

- What happened at the Big Bang
- What happens when Neutron Stars collide

LIGO proved that Gravitational Waves, another of Einstein's audacious predictions made 100 years earlier, do exist. Gravitational Lensing and GPS prove that, unless we take account for the Space and Time aspects of Space-time, our new technologies will not work.

11   How to Win a Nobel Prize

Throughout this book I have identified a number of scientists whom have been awarded the Nobel Prize for their contribution to our knowledge of Space-time. In this chapter I am going to consider just how difficult this was and challenge you to see if you could now get involved. Hopefully, this will show you how close and yet how far away the Nobel Prize is for anybody. (My second book 'Quantum Physics for Absolute Beginners' tackles this challenge using Quantum Physics as the facilitator.)

You now need to try and make sense of your new knowledge in a way that nobody has done before. Don't worry; I am still here to help. Our journey has taken some deliberate diversions along its way so let us remind ourselves of the key things that we have learned.

1. **Let's take stock of what we know about Real v Perceived**

We have shown that Gravity is not Real; it is Perceived:

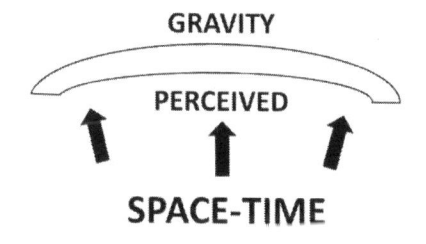

GRAVITY

PERCEIVED

SPACE-TIME

REAL

It is the curvature of Space-time which creates the force of Gravity that we perceive.

It is Space-time that is Real. Space-time is experienced the same by everybody in every location (i.e. **absolute**).

2. What double acts have we discovered?

We have identified some double acts that are equivalent. We have encountered at least two of these when we study motion:

CONSTANT SPEED ⇄ STANDING STILL

GRAVITY ⇄ ACCELERATION

As far as the laws of physics are concerned, moving at Constant Speed and Standing Still are the same.

Gravity and Acceleration are also equivalent. If you close your eyes in either of these situations they appear to be the same thing.

Other double acts cannot exist within our universe without the other. This is true of Matter and Light as well as Space and Time:

MATTER ⇄ LIGHT

SPACE ⇄ TIME

Energy and Mass are also equivalent; they may be interchanged without contravening any Law of Physics (using a conversion factor of c^2).

Something similar happens with Matter and Anti-matter pairs – once we remove charge one is almost the same as the other.

ENERGY ⇄ MASS   MATTER ⇄ ANTI-MATTER

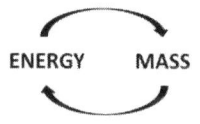

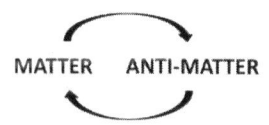

Some fields also appear to be equivalent; an Electric Field will induce a Magnetic Field and vice versa:

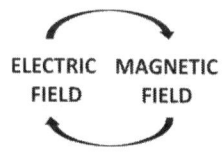

ELECTRIC FIELD ⇄ MAGNETIC FIELD

Then we have Space Speed and Time Speed:

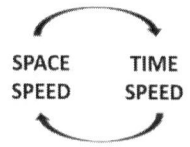

SPACE SPEED ⇄ TIME SPEED

You could argue that this fits alongside all of the five double acts above it (the ones governing motion appear to be different). In fact, you could probably make an argument for categorising those six double acts in all sorts of different ways.

If you can find a new basis for describing or categorising these 6 double acts this might be addressing a very important question, one to catch the imagination of the scientific community. This would be an excellent way to kick start your quest.

Have you spotted any more double acts? This might instigate a new line of questioning.

When you are ready we need to confront the main challenge.

3. If Space and Time are Perceived then what is Real?

This is the first question posed in Chapter 1. It was originally asked by Einstein in 1905 and we had a diagram to explain it:

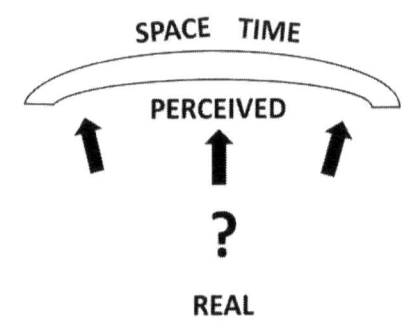

SPACE   TIME

PERCEIVED

⬆ ⬆ ⬆

?

REAL

Ten years later Einstein answered his own question and showed that it is **Space-time** that is Real:

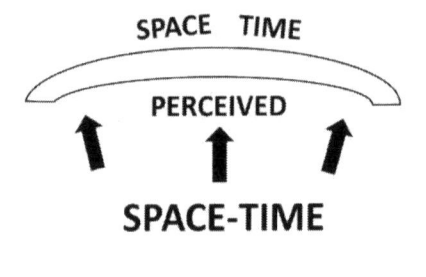

SPACE   TIME

PERCEIVED

⬆ ⬆ ⬆

SPACE-TIME

REAL

Space and Time really are just Perceived (like Gravity). It is **Space-time** that is Real.

So within this book we have answered the first question. But what about the other two:

a. What changed to make Space-time Real?
b. When did this happen?

4. What changed to make Space-time Real?

We now know that Space-time can reveal itself in different ways so we could add a similar question which is 'What caused Space-time to transition?' This question is just as good.

We know that physical Laws appear after something significant changes. Scientists say that Symmetry gets Broken. We have identified a few contenders:

- The universe cooling might have been a catalyst
- Alternatively, it could have been the process of expansion and reduction of density (at the Big Bang or as part of a Cyclic Event) that sparked a change
- There might have been a movement from many dimensions (maybe 10 or more) to the four dimensions we see today
- Another option is that two or three catalyst items above may have happened simultaneously

5. When did this happen?

In this book we have identified three occasions when this might have happened. We know that Positional Symmetry was broken at least three times during the evolution of the universe:

- The Big Bang
- Inflation
- The Plasma period that led up to the 'Light Age'
- Also, something happened six billion years ago to make the universe expand again. Could that have been another Break in Symmetry?

6. What is the Space-time Speed evidence?

a. We know that there is a tight relationship between Space and Time and they are linked by the Speed of Light
b. There are also a growing number of scientists who support the idea of a Cyclic Universe - as our current universe expands and gets less dense with negligible mass identifiable anywhere this may have been the origin of a **Big Bounce**; everything starts again as if there were another Big Bang
c. The SPACE-time Speed environment for Inflation would support this theory
d. However, the Space-TIME Speed environment that supports life is very different. If you accept the theory of Cyclic Universes you have to explain why the universe oscillates between different environments

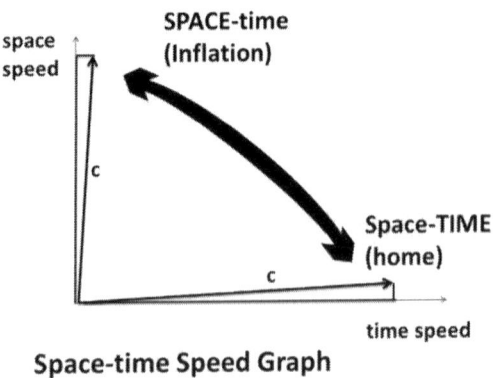

Space-time Speed Graph

Do we live in our Space-TIME environment because of the presence of limited mass (or energy) in our Galaxy or Solar System?

Is there a different Space-time environment between galaxies?

Would the Law of Disorder make Space-time itself less orderly; would it separate Space and Time?

7. Is there another double act at work here?

We have already seen that the Universe Heat Death could lead to a SPACE-time Inflationary environment. During Heat Death if Time were to suddenly reduce to an infinitesimal level there would then be the SPACE-time environment to support Big Bounce Inflation. We appear to have:

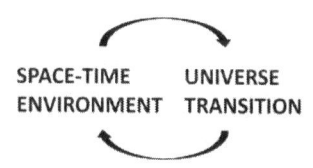

SPACE-TIME ENVIRONMENT ⇄ **UNIVERSE TRANSITION**

If you can explain why the transition could take place then you would be on to something new.

If you could provide a full explanation for Big Bounce Inflation (without having to contradict the Universal Speed Limit) then it would be very inconsistent if you did not receive a Nobel Prize nomination.

You now know the key concepts, required to make a new breakthrough in Physics but you will need to complete a 'knowledge' jigsaw puzzle and find some new connections.

8. How concise can we get?

You are just one sentence away from winning the Nobel Prize:

> **The universe will transition into a new Space-time environment when [A] and this will happen [B] which we can prove by [C]**

We have some options for **A** which are:
- It cools rapidly
- Its density is quickly reduced
- It loses dimensions

And we also have options for **B**:
- at the Big Bang
- at a Big Bounce
- during a Plasma period

C is harder but you will need this to substantiate **A** and **B**. Consider the following string of relationships:

> **MASS -> ENERGY -> FREQUENCY INCREASE -> SPACE-time**

(NB. Big Bounce Inflation arguably is a Space-time environment)

If you can explain the connections then you are getting very close. Remember that Space-time is a field moving at the Universal Speed Limit so how can we link 'FREQUENCY INCREASE' to SPACE-time?

9. If you don't have an answer, do you have a Question?

Use your imagination to come up with something new. Most of the Nobel Prize winning scientists named in this book started off with a question. If you can find a new question then you will be on the right track and, I believe, there are still many important questions out there.

Once you know a little about the reality of your world you can ask new questions. If I tell you that the only difference between two photons is their frequency (and wavelength, of course) try to answer the question 'Where does colour come from?' (See Notes 8 if you need to).

Asking a question that nobody has asked before is the most likely route to finding something new. This is the central theme of this book; question everything and see if you can be the first to square some of the circles we currently have in Physics.

If you believe that you have a new question add it to the Quora website. The scientists there will quickly confirm if it is really new.

10. What are the options for winning a Nobel Prize?

There are six ways to win a Nobel Prize. Winning the Peace Prize is something that happens to people. It is not something that can be easily planned. Similarly, literary critics will argue that the Literature Prize is something that just happens to you.

That leaves the science based awards. These comprise Physics, Chemistry, Medicine and Economics (although to be precise this is the Sveriges Riksbank Prize in memory of Alfred Nobel). The oldest science and most prestigious of these is Physics and that is what you now know something about.

11. What is the process for winning a Nobel Prize?

Within theoretical physics anybody that wins a Nobel Prize has to adhere to the following process:

a. Discover a new Theory
b. Show that there is an experiment (or observation) that would prove your Theory

c. Get your Theory seen by publishing it in a scientific journal (the Physical Review is the most prestigious)
d. Get somebody to confirm in a journal that the suggested experiment works
e. Wait whilst the Nobel Foundation determines a short list (there is a lot of waiting at this stage)

Once they accept that your contribution has been the 'most worthy' from their short list you will receive an invitation from the Nobel Foundation to visit Stockholm and collect your Nobel Prize from the reigning monarch:

f. You never know when this is going to happen but, if you have followed the instructions above, you never know …
g. Your success will be made public during the first week in October (you will be told about 30 minutes before it is made public)
h. Try to look modest whilst the science community comes to terms with the fact that you discovered this before any of them
i. Keep your Decembers free for about 15 years (Einstein had to wait 16 years)
j. When you get there you may mention my name, although I expect you will, by then be too famous!
k. If you do tell people that it was my book that inspired your discovery then may I take this opportunity, in advance, to express my gratitude.
l. And, well done!

12. Can you really win?

Alfred Nobel, b.1833 – d.1896, left his wealth to establish the Nobel Prize in 1895. According to the rules of the institution, anybody (i.e. not just scientists) really can win the Nobel Prize, but nobody says that it will be easy!

Conclusion

1. Give me some space

If somebody asks you to give them some space it probably means that your relationship isn't going as well as it might. However, for the purposes of this book, most importantly you would know what they meant. You are prepared to accept that different people can have their own space.

In the same way that somebody can have their own Space, they can also have their own Time. In fact, we all have our own Time. All Space and Time is our personal property.

Because we all live on this planet together, moving at the same speed through the cosmos together in the same gravitational field we all experience the same Space and Time. But this is an illusion. As soon as one of us leaves our planet the Space and Time experienced will change.

Knowing this new fact (it has been proved by GPS) isn't going to change your life but it may change the way you view our place in our universe. After all, that is what we all want to know isn't it?

Finally, you may not have managed to figure out a new theory but let's look at what you have achieved:

1. You now know something about reality (which has been proven by scientists). We have worked our way through over 80 sections of which many have clear questions; some have clear answers, and if there are none you now know why.
2. You can see how the Big Bang probably happened, why previously there may have been a time when there was no Time and why the Universe rapidly expanded during the first second.
3. You are also aware that there may be alternative explanations for our universe and if you can take the Space-time argument to a conclusion then your explanation for a Big Bounce could be the new theory that we are waiting for.
4. If you have followed the trail through this book you will have learnt new facts and you will have tried to apply these to the conundrums outlined; this is what science is about and this is why it is exciting.
5. Because you now understand the 'Scientific Method' you can win arguments against people who have long held but wrong ideas which they cannot prove such as astrology and homeopathy.
6. You now recognise that popular science isn't just for scientists and that you have entered a new world of discovery which is more amazing than anything that you have ever seen in science fiction.

If you haven't already done so, take a look at the Notes. All of the arithmetic is at an elementary level but an understanding of the mathematics can sometimes give you a deeper comprehension of nature.

If you are concerned about some of the detail within the chapters then please take a look at 'Holes in the Truth' before you complain. If you have any further questions you are welcome to contact me on Twitter at @TonyGoldsmith10. If I cannot answer your question I can probably suggest someone who can.

When I set out on this project my vision was to write science that could be understood by anybody with the ability to read. Consequently, I have spent a lot of time working and reworking sections that needed to be improved and I am sure I have not finished yet. If you have any good ideas for improving this book then please also contact me at the above address.

At this point within the book you will know more about Spacetime than most physicists knew just 30 years ago. You may not, as yet, have found any new connections, but whatever the outcome….

Welcome to the Science Community

Notes

This section is referenced from the Chapters:

1. Newton's Laws

a. In Chapter 3 we refer to Newton's Laws of Motion so here is a reminder of what they are.
b. In 1687 Newton famously gave us three laws of motion:
    1. An object will remain static or at constant speed unless acted upon by a force (aka Law of Inertia)
    2. The Force applied to an object is equal to its Mass multiplied by its Acceleration (F=ma)
    3. For every action there is an equal and opposite reaction.

2. Speed of Light with examples

a. In Chapter 3 we discuss the implications for the Speed of Light being constant. The upshot of this statement is that both Space and Time must be changing. This is central to this book and I am keen to give you as many perspectives on this statement as I can. Here is another.
b. If you have broken out to this Note then you are either having a problem with terminology or the concept of an equation. Either way we can fix it.
c. Let's write down the Speed of Light as an equation:
    Speed of Light = $\dfrac{\textbf{Distance covered}}{\textbf{Time taken}} = \dfrac{\textbf{300,000 km}}{\textbf{1 sec}}$
    This is what a man in the rocket would measure.
d. Now let's assume that, coincidentally, the distance from the rocket to the man on Earth is 300,010 km. The man on Earth will see the photon travel for 1.00003 secs because:
    $\dfrac{\textbf{Distance covered}}{\textbf{Time taken}} = \dfrac{\textbf{300,010}}{\textbf{1.00003}}$ **= 300,000 km/sec (always)**

e. Both the men observe the same event but each sees the photon travel a slightly different distance in different times because they are in different places moving at different speeds (i.e. different reference frames).
f. For light, when the perceived distance (Space) changes the Time must change too.
g. I know it is strange, but it's true!

3. Length contraction calculation

a. In Chapter 3 we state that as you get closer to the Speed of Light the length of an object contracts and time slows down. The effect of speed on time is fully covered elsewhere in this book so this section looks at what length contraction really means and we show the maths which substantiate a 50% change.
b. We posed the question that if the pole vaulter's pole shortens by half (5m to 2.5m) how fast must he run.
c. The answer is given by the equation:

Moving length = Rest length $\times \sqrt{1 - \dfrac{v^2}{c^2}}$

Hence when: $\dfrac{\textbf{Moving length}}{\textbf{Rest length}} = 0.5 = \sqrt{1 - \dfrac{v^2}{c^2}}$

v must be **86%c**.

d. NB. This equation was originally written as:

Lorentz factor $= \sqrt{1 - \dfrac{v1^2}{v2^2}}$

Where v1 and v2 are velocities for reference frames 1 and 2

Einstein just changed v2 into c and removed the v1 subscript. It can be as easy as that! All of the above equations are still known as Lorentz Transformations though.

e. For those of you that are wondering what happened to the mass of the pole, it hasn't magically disappeared. No, instead it has increased according to the equation:

$$\text{Rest mass} = \text{Moving mass} \times \sqrt{1 - \frac{v^2}{c^2}}$$

f. Instead of decreasing the Moving Mass will increase!

4. **Speed of Space-time**

We know that Space-time includes components of Space and Time such that they complement each other. We know that when Space is dominant, Time takes a low profile but what are the factors that make Space dominant. Brian Greene used Space-time Speed as the factor for analysis and, for 'Absolute Beginners' I think this works really well.

a. In Chapter 5 we introduce the speed relationship between Space and Time and Greene suggests that this can be represented on a right angled graph
b. This analysis is a new line in Physics and, because of this I have taken care not to show any units (for Space or Time speed)
c. I am also aware that the relationship might not be based precisely upon this shape of graph (i.e. the axes could cross at 60° instead of 90°) but, I am dealing with principles and a right angled graph is much easier to work with
d. Because Space speed and Time speed are relative to their locations I have been careful to use ratios for Space speed and Time speed and these are the basis of predictions for two different types of Space-time environments
e. The implications for these environments are further discussed in Chapters 7 and 8.

Maths behind the Space-time Speed Graph

f. The figure below is an extended version of the figure shown in Chapter 5 which displays the relationship between space speed and time speed on a Space-time Graph. Note that this is not a Distance (Space) v Time Graph.

g. I have taken two scenarios, both of which have Space-time moving at the Speed of Light (c).

h. The first is a real life scenario where we travel through space at 14 km/sec relative to Earth (this is about one twenty-thousandth of the speed of light). You may remember that this was the speed of New Horizons when it passed by Pluto in 2015 (the fastest man made object ever).

i. The second scenario turns that on its head and sees what would happen if we could travel through space faster than time.

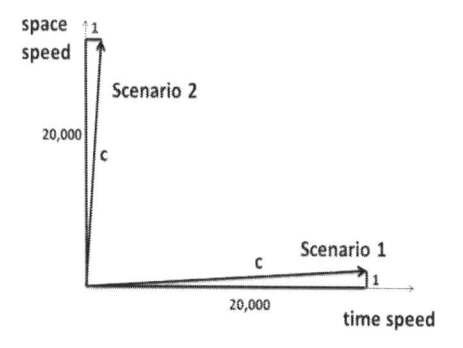

Space-time Speed Graph

j. Both of the scenarios described are based upon a right angled triangle so we could use Pythagoras's Theorem to calculate the respective sizes of the sides.

k. Pythagoras's Theorem states that the square on the hypotenuse (c in our examples) is equal to the sum of the squares on the other two sides. If you want to you can work out how much of c is moving space wards and how much is time wards. However, for these sizes there is an easier way:
l. If the space-wards line is 1 cm (as shown in scenario 1) then we would need a bottom line which is 200 metres long to keep this to scale. Instead, we can reasonably assume that the bottom line is near enough the same as 'c' (say 99.99% c at least).

For Scenario 1: Space-TIME

m. We cannot be certain of where our Space-time fits into the universal order but we have some clues; it is possible for us to travel faster through space but not easy to go slower; we know that time can slow down (with higher velocity or gravity) but it is hard to speed it up
n. These clues suggest that we live at the edge of Scenario 1
o. In the diagram the amount moving space wards is approximately one twenty thousandth of c and that equals about 0.005% times the speed of light.
p. At this speed the space contraction is less than one tenth of a percent. This is not noticeable using human senses
q. Nobody knows what effect this might have on the human body (should you be on New Horizons) but work is proceeding to understand this better
r. The amount moving time wards is almost the same as c (over 99.99% times the speed of light) and the amount that time would slow at that speed is also less than one tenth of a percent
s. Consequently, even though we can travel at 14 km/sec we would not notice any big differences in the relationship between space and time
t. This scenario is the world that we live in.

For Scenario 2: SPACE-time

u. In the second scenario we have the same sized triangle but it is turned on its head (or through 90° if you prefer)
v. We move space-wards at 99.99% of the Speed of Light
w. At this speed the size of the rocket will undergo **length contraction** and will reduce by over 224 times
x. Simultaneously the time-wards motion is now 0.005% of the Speed of Light
y. This means that the accompanying **time dilation** time will slow down by over 224 times
z. We would not recognise this world but it might have been the foundation for an Inflation type of expansion during a cyclic universe (and this could have taken much longer than that allowed for in the current Inflation Theory).

5. Powers

a. From Chapter 3 we start to use some large numbers. To avoid having to write them all down in full (i.e. 10^{26} is 100000000000000000000000000) it is time to clarify the scientific notation that is used
b. The use of powers is a useful tool for viewing big numbers
c. The simplest power is a square (i.e. $4^2 = 4 \times 4 = 16$)
d. A cube is the number multiplied three times (i.e. $4^3 = 4 \times 4 \times 4 = 64$)
e. The most useful base for our powers is 10:
    i. 10 to the power of 2 is 100 (i.e. 10 x 10)
    ii. 10 to the power of 3 is 1000 (i.e. 10 x 10 x10)
    iii. 10 to the power of 1 is 10
    iv. 10 to the power of 0 is 1 (it just is).
f. Note that the powers of 10 are the same as the number of zeros in the final number.
g. When we multiply two numbers we can just add the powers:
    For example: $10^2 \times 10^3 = 10^5$
h. When we divide we just subtract:
    For example: $10^6 / 10^3 = 10^3$

i. This is also the basis of Logarithms which are just powers:
   For example   Log 2 (base 10) is 100.
   In the same way that $10^2 = 100$
j. When we divide a powered number we get negative powers:
   For example: $\dfrac{1}{10} = 10^{-1}$

   $\dfrac{1}{10^3} = 10^{-3}$

6. Exponential Graphs

a. In Chapter 6 we extend our view of the universe and try to consider 13.8 billion years on one diagram. To do this we need to understand the concept of exponential units and how they can be applied to graphs
b. An Exponential Graph is shown below:

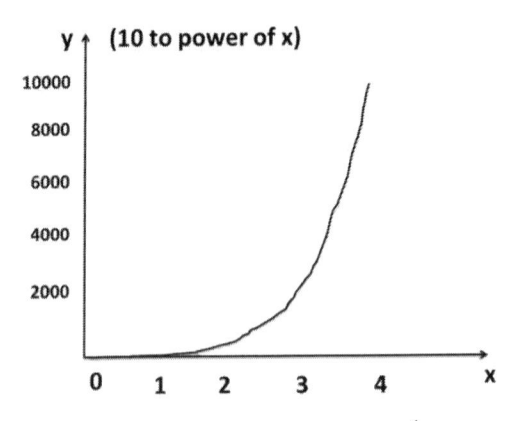

y ↑ (10 to power of x)

- 10000
- 8000
- 6000
- 4000
- 2000

0   1   2   3   4   → x

i. For x=1 the y value is 10 (10^1)
ii. For x=2 the y value is 100 (10^2)
iii. For x=3 the y value is 1000 (10^3)
iv. For x=4 the y value is 10000 (10^4)

c. The y value gets big very quickly (10 times bigger at each stage)

d. We need to use this technique when we want to see very big and very small numbers at the same time
e. Note that this graph is a different shape to that shown in Chapter 6 for Inflation. A pure exponential graph is a steady increase

7. How does our Sun work?

a. In Chapter 6 we consider how our Sun works
b. Our Sun is made of hydrogen; this is the simplest atom that exists (one proton and one electron) and most of it was made during the early Plasma period
c. When hydrogen atoms fuse together they eventually manufacture helium. This takes place during a three stage process) which has waste products which include positrons (yes, anti-matter) and Gamma Rays (photons). Stable helium atoms are placed into storage at the centre of the Sun
d. The photons created spend the next million years (or so) fighting their way to the surface of the Sun. They lose a lot of energy on their internal journey but they then escape into the solar system as visible light. From the surface of the Sun it takes about 8 minutes (Earth time) for one to reach us on Earth
e. The electromagnetic energy (light and heat) that we receive from the Sun sustains all life on Earth

8. Where does colour come from?

a. In Chapter 11 we ask this question
b. This book encourages you to find some new questions and uses this as an example of one that you could make up
c. You have all of the information required to answer this question
d. You know that it is a photon that hits your eye and you know that the only properties inherent in a photon are frequency and wavelength (photons do not have colour)

e. Therefore, when the photon interacts with your retina (like a key finding the right lock) the only thing that happens next is an electric spark is sent to your brain
f. It is your brain that adds the colour and the form and the perspective and everything that we see
g. Your brain does something similar with messages from all of your senses
h. You can see that the answer can be deduced from the basic knowledge you now have but, like all good questions, it leads on to something else; in this case the realisation of just how amazing our brains are!

Holes in the Truth

This book has been designed to focus upon those Space-time facts which usefully lead to an understanding of reality which is supported by mainstream physics. This means that, during this journey, I have chosen not to include every detail of every explanation. Consequently, there may have been a few occasions in this book when I have been economical with the truth. It is now time to tidy up:

- Whereas Maxwell did create the equations necessary to calculate the Speed of Light it is unlikely that he had all the data required to complete the job in 1865 (e.g. Displacement Current)
- OK, there is a distance v time graph within the book but it is not there to calculate speed – I just needed to show how unbelievable Inflation is
- When examining atoms we have focused upon the three main particles of protons, neutrons and electrons. We now know that protons and neutrons are made out of various types of quarks but I do not see how this additional level of detail would help improve the focus of this book
- My photon diagram might be a little bigger than it should be (arguably it should be one wavelength in size) but I just want you to be sure of what it is when you see one
- The Earth does not orbit the Sun in circular motion. It is an elliptical orbit (just) which means that during summer and winter the orbital speed increases very slightly. It doesn't, however, detract from the point I was making about constant speed (it's a tiny difference) but it is the reason why a sundial is not always accurate
- Also, electrons do not just orbit the nucleus in an atom. They may be the smallest particle currently known but they exist in a comparatively large field. What they do is quite fascinating but you will need to read a my book 'Quantum Physics for Absolute Beginners' to understand this

- Because of the above disclosure I might still struggle to convince you that 99% of an atom is empty space – clearly there is something else going on
- An ion is not just a chunk fallen off an atom – it is a chunk that has a positive or negative charge (but, other than anti-matter and dimensions condensing, I have managed to avoid talking about charge and I don't want to start here; this is a book about Space-time)
- As well as frequency and wavelength photons do exhibit other features (such as polarisation) but none of this detracts from the central message which is that they are very basic building blocks of nature and most of our personal perception of the world that we live in is done in our brains
- Throughout this book we only consider the speed of light in air (or to more precise its speed in a vacuum which does not have stray molecules in the way of passing photons). Its speed does change in other substances; in water the speed of light is about $0.75 \times c$ and in glass it is about $0.67 \times c$
- When discussing entropy we said that things always get more disorderly – well, according to the known Laws of Physics it is possible for an uninfluenced system (i.e. broken egg) to spontaneously become more orderly (i.e. whole again) but the probability of this happening is so tiny that it is highly unlikely that anybody will witness this event during the life of our universe (or subsequent ones)
- As an object gets closer to the Speed of Light we said that its mass is increased. This is true but to understand this better you need to understand that Mass is the ease with which something is moved (i.e. inertia). As we get closer to the Speed of Light an object will rapidly reduce its ability to accelerate. It will not necessarily become massive (according to the dictionary definition) but the immovable effects will be the same
- Not everything in our universe moves anti-clockwise; Uranus spins clockwise but that is because it is believed that another planet smashed into it during the early years of the Solar

System (see 'Why is Uranus Upside Down?' in my Further Reading list)
- When two hydrogen atoms fuse to create light helium ($^{2}_{2}He$), as well as an emitted electron a neutrino is also released (but I haven't seen a good reason to explain what a neutrino is so I hope I can be forgiven for ignoring it; this book is about Space-time)
- It may be that 'all other' elements are not manufactured during a Supernova event. The latest hypotheses suggest that the heavier elements (like Uranium) may have been manufactured during Neutron Star collisions
- Many scientists would argue that the Space speed v Time speed diagrams are too simplistic (and do not take account of Minkowski metrics etc.) - my response is that this is a book for 'Absolute Beginners' and I offer no apology for trying to simplify everything!

Further Reading

I am responsible for some of the perspective and terminology in this book. However, none of the scientific theories are my own. Just about all of the intellectual content came from the publications listed below which I have also reviewed for the reader.

1. Introduction to Special Relativity (1968) – Robert Resnick
   All of the books I had at University are now very old and I did not read much of any of them (I was not a good student). I did, however, read this book more than once and I was completely enthralled by the idea that space and time are not absolute. What else could I not rely upon?
2. A Brief History of Time (1988) – Stephen W. Hawking
   I did read all of this when it first came out (it was a present). Clearly, it was the catalyst that led to me reading a further 80 (and counting) or so books on physics and reality. However, when I first read 'A Brief History ...' it seemed to raise more questions than answers; I'm not sure we should blame the author though – it was probably just me!
3. From Eternity to Here (2010) – Sean Carroll
   This was the last book that I read before undertaking this project. As such it is the inspiration for this book. I first read it some time ago but missed out the first sections and went straight to the bits I thought would be new to me. I never finished it but I have picked it up again recently and read it from the beginning. Clearly, I hadn't understood the early chapters well enough but the author quickly fixed that. He has a gift for explaining mind boggling principles in easy to understand terms. I hope my book does something similar for an uninitiated audience.
4. The Fabric of the Cosmos (2004) – Brian Greene
   Classic popular science by a legendary author
   Given my focus upon Space-time Speed you probably ought to read this next.

5. Einstein His Life and Universe (2007) – Walter Isaacson
   This is also a good place to start. You get to really understand Einstein's life and times and what drove him to discover new models of reality. You also get a gentle introduction to his famous theories.
6. Love and Math (2013) – Edward Frenkel
   This is a great book on so many different levels (including politics). Treat yourself.
7. Why is Uranus upside down (2007) – Fred Watson
   If you ever look up to the night sky and wonder at the starry panorama then this book is for you. It takes all the frequently asked questions (literarily) and provides easily understandable answers. Some explanations, like why we see a large moon near the horizon, will amaze you.
8. 1001 Wonders of the Universe (2011) – Piers Bizony
   If you want to *see* all of the wonders of the universe this book has them all.
9. Wonders of the Universe (2011) – Brian Cox and Andrew Cohen
   Excellent book and excellent TV series
10. Wonders of the Solar System (2010) – Brian Cox and Andrew Cohen
    ditto
11. The Particle at the End of the Universe (2012) – Sean Carroll
    It all seems old hat now but at one time it was like the only scientific research going on was the hunt for the Higgs Boson.
12. The Nature of Space and time (2015) – Stephen Hawking & Roger Penrose
    Requires a knowledge of mathematics
13. Farewell to Reality (2013) – Jim Baggott
    A reasonable rant.
14. What if Einstein was Wrong (2013) – Various
    I was disappointed with this; very little on Einstein
15. New Scientist magazine – circa 1995 to present
    Would anybody like to buy 22 years of back copies?
16. Gravity's Engines (2012) – Caleb Scharf
17. The Book of Universes (2011) – John D Barrow

18. Cycles of Time (2010) – Roger Penrose
    So no single Big Bang then
19. Why does $E=mc^2$ (2009) – Brian Cox and Jeff Forshaw
20. 13 things that don't make sense (2009) – Michael Brooks
21. We need to talk about Kelvin (2009) – Marcus Chown
    Brilliant author
22. Light Years (2008) – Brian Clegg
23. Why Beauty is Truth (2007) – Ian Stewart
    More maths than physics but a great read
24. The Never Ending Days of Being Dead (2007) – Marcus Chown
25. The Goldillocks Enigma (2006) – Paul Davies
    Another of my favourite authors
26. What we believe but cannot prove (2005) – Various
27. The Infinite Book (2005) – John D Barrow
28. The Road to Reality (2004) – Roger Penrose
    This is the daddy of all popular physics books. This one has everything we know up to 2004. I promised myself I would read it when I retired.
29. Critical Mass (2004) – Philip Ball
30. A Brief History of Infinity (2003) – Brian Clegg
31. Our Cosmic Habitat (2001) – Martin Rees
32. About Time (1995) – Paul Davies
33. Surely you're joking Mr Feynman (1985) - Richard Feynman
34. Six Easy Pieces (1995) – Richard Feynman
35. Black Holes and Baby Universes (1993) – Stephen Hawking
    Yes, I went through a period when I got all of Stephen Hawking's books. I still have these.
36. The Universe in a Nutshell (2001) – Stephen Hawking
37. On the Shoulders of Giants (2004) – Stephen Hawking
38. A Briefer History of Time (2005) - Stephen Hawking
39. The Tao of Physics (1975) – Fritjof Capra
    We thought there might be a spiritual connection!
40. Straight and Crooked Thinking (1953) – Robert Thouless
    I have always loved this book. It has taught me so much.

Please excuse me if I have not read every publication in this field. There is a finite amount of concentration time in every person's life!

Index

The following list shows when each of these topics is introduced according to their chapter and section numbers:

Acceleration	5:2
Anti-matter - in Vacuum Space	2:8
Big Bang	6:3
Big Bounce	7:4
Blue shift	6:2
Cause and Effect	3:10
Constant speed	3:1
Cosmic Microwave Background	6:8
Cosmological Constant	6:0
Classical Relativity	3:2
Copernicus	6:5
Dimensions	8:5
Disorder	4:3
Doppler Effect	6:2
Double Acts	1:1
Dwarf - White & Black	7:2
Eddington, Arthur	5:3
Einstein	
- Introduction	3:0
- Special Relativity	3:4
- General Relativity	5:2
Electromagnetic	
- Spectrum	2:4
- Propagation	2:6
Electron	
- Position	2:1
Emergent	1:1
Energy	3:7
Faraday, Michael	2:5
Field	2:5
Flatness problem	6:5
Formalont, Ed	5:5
Fundamental	1:1
Galaxies	7:0

Galileo	3:2
Graph	2:7
Gravity	
- Equivalence	5:2
- Speed	5:4
Gravitational	
- Lensing	10:2
- Waves	10:3
Guth, Alan	6:7
GPS, trilateration	10:1
Heat Death	8:7
Helium atom	2:1
Hoyle, Fred	6:1
Horizon Problem	6:6
Hubble, Edwin	6:1
Hubble Space Telescope	6:2
Inflation (after Big Bang)	6:7
Invariance	8:1
Kopeikin, Sergei	5:5
Length Contraction	3:6
Light	
- Definition	2:2
- Constituent parts	2:4
- Age	6:8
LIGO	10:3
Mass	3:7
Matter	2:1
Maxwell, James Clerk	2:6
Michelson, Albert	10:3
Milky Way	7:1
Morley, Edward	10:3
Multiverse	7:4
Nebulae	7:0
Neutron	2:1
New Horizons	9:4
Newton, Isaac	3:0
Nobel, Alfred	11:12
Nobel Foundation	11:11
Omega, density	8:4
Penrose, Roger	6:3
Perceived (books definition)	1:1

Perimutter	6:1
Photon	2:3
Plasma Period	6:8
Positron	2:8
Proton	2:1
Proxima Centauri	9:4
Quasar	5:5
Real (books definition)	1:1
Red Giant	7:1
Red shift	6:2
Riess	6:1
Second Law Thermodynamics	4:3
Schmidt	6:1
Scientific Method	11:11
Space - Curvature	5:1
Space-time	
- Absolute	5:7
- Speed	5:8
- Space and Time Speed	5:9
- Catalyst	8:3
- Curvature	9:3
- Shortest distance	9:9
- Longest time	9:9
Speed of light	
- Measurement	2:2
- Calculation	2:6
- Faster than	6:7
Solar System	7:1
Sonic Boom	6:2
Star	
- Medium size	7:1
- Small size	7:2
- Large size	7:2
Stardust	7:2
Sun	7:1
Symmetry	
- Types	8:1
- Breaking	8:1
- Boost	8:1
Supernova	7:2

Time
- Definitions   4:1
- Measurement   4:5
- Dilation   3:6
- Arrow   4:4
- Travel   9:5

Universal Speed Limit 3:8

Universe
- Static   6:0
- Expanding   6:1
- Accelerating   6:2
- Cyclic   6:3

Vacuum Space   2:9
Virtual Particles   2:9
Voyager 1   7:0

About the Author

Tony Goldsmith graduated in Physics at Leicester University in 1974 and then spent a full working career in IT. He has returned to science over the last 25 years to try and satisfy his curiosity for what is real and what is not. He has also taught Physics; briefly in a secondary school and, more recently, as a private tutor. Consequently, whenever he encounters a new topic his first reaction is to figure out how best to teach it (without equations). He hopes that this new perspective will help to explain the science landscape for people with little or no science background. It is for people like you that he has written this book.

Printed in Great Britain
by Amazon